Field Studies, **10**, (2001) 1-35

A KEY TO THE FAMILIES OF BRITISH BUGS (INSECTA, HEMIPTERA)

DENNIS UNWIN
Museum of Zoology, University of Cambridge, Downing Street, Cambridge CB2 3EJ

ABSTRACT

This illustrated dichotomous key is not intended for experts, who have no need for it, and is limited in scope to what can be achieved by examining unprepared specimens through a stereo dissecting microscope of moderate power (up to x50). Nevertheless, the user will be able to identify the adults of most British bugs to family level; but there are two groups, aphids (Aphidoidea) and scale insects (Coccoidea) where more sophisticated techniques, beyond the scope of this guide, will have to be used. As for other AIDGAP keys, the text and illustrations were extensively tested during preparation; this, the first published edition, has been produced in the light of feedback from the testers.

INTRODUCTION

The first stage in the identification of an invertebrate is to discover what sort of animal it is, and that means placing it first in a phylum, then a class and finally an order. This may seem to be a simple matter, but there are some very atypical animals in every order and it may not always be a simple task. A guide to an enormous group, such as the invertebrates, inevitably has to generalise and a totally watertight key would be a very large document if it dealt with all the oddities. Nevertheless, there are many over-simplified keys that take one to the order which will fail all too often. A very good compromise is the key to the major groups of British terrestrial invertebrates by Tilling (1987).

Having arrived at the order, the next important step is to discover the family, which is the purpose of this key as far as the bugs are concerned. Once the family is known, one can then look for a key to take the identification to species level. Some bug families have modern keys, although the process of identification may not be simple; the bugs are a difficult group and some families are more difficult than others. In other cases, older keys have to be used, some of which can only be obtained from libraries. Here one has to be aware of changes in classification, as a family in the modern sense may not be quite the same as in an older text. There will also be changes to some of the names, for a variety of reasons. To help in these matters there is a checklist (Kloet & Hincks, 1964) although this is now quite old and many changes have taken place since it was published. At the time of writing, a new checklist is in preparation. In the meantime, the exact number of species in a family seems to depend on whom one asks; when this information is given it should be taken as approximate. Since the only checklist available when this key was written was rather old, the classification is based on that used by Dolling (1991). A table of the families recognised in this key is given on page 8. Many of the works on bug families available at present will use an older family concept, and where possible this will be indicated in the text.

This key is not intended for experts, who have no need for it, but is aimed at anyone who needs to find out the family to which a bug belongs and is not familiar with the group. I have attempted to avoid unnecessary jargon but, to take the identification further, some familiarity with the descriptive terms is desirable. These are often not explained in the more specialised keys. When such terms are used in this key they will be explained. In order to simplify the use of the key, it has been necessary to bring out some families in more than one place, but where this has been done, there will be a note to the effect with the family name.

This guide is limited to what can be achieved by examining unprepared specimens with a stereo dissecting microscope of moderate power (up to x50). This will enable the user to identify most of the bugs to family level, but there are two groups where more sophisticated techniques need to be used: these are the aphids (Aphidoidea) and the scale insects (Coccoidea). Both groups require specimens to be prepared and slide mounted, so that they can be examined under a high power compound microscope, and such techniques are beyond the scope of this guide. The aphids are a difficult group in which the taxonomy is somewhat unstable and although keys do exist to many of the species, they are of necessity difficult to use. In the circumstances I have decided to list the characteristics of each of the main families rather than attempt to construct a key that would not be reliable. Nevertheless, many of the aphids that will be encountered can still be placed in a family with reasonable accuracy by reference to the family descriptions. A family key is given by Dolling (1991), but it is intended for use with slide mounted specimens.

The other group not keyed out to family are the scale insects (Coccoidea) which are very strange animals indeed. This is a difficult group, and the information available is limited, but some help has been given. Dolling (1991) provides a family key, which again will require slide preparation in many cases. I hope the experts will forgive this pragmatic approach, but the objective is to give as much help as I can to as many people as possible, even if this means a somewhat uneven treatment.

What is a bug?

The word 'bug' has a variety of meanings which may range from an arthropod, or even a micro-organism, to a problem in a computer program; but to the entomologist a bug is a member of the order Hemiptera. For this reason the Hemiptera are called 'true bugs', although this term is more often reserved for the Heteroptera, one of the sub-orders of Hemiptera. This can be very confusing, but in this key the term 'bug' means a member of the order Hemiptera.

Hemipterans are a very diverse order of insects that range from less than a millimetre to forty millimetres in length. They have mouthparts in the form of a rostrum, sometimes referred to as a beak, which is held under the insect, when not in use, pointing rearwards and downwards but never forwards (Fig. 1a). The rostrum is divided into segments, usually three or four in number. Bugs feed on fluids, which may be of plant or animal origin. There are in excess of 1600 species in the British Isles, which are grouped into three sub-orders; Heteroptera, Auchenorrhyncha and Sternorrhyncha.

There are just over 500 species of heteropteran bugs in the British Isles, and the group includes the predatory and bloodsucking bugs, all the true water bugs and the plant bugs. Most mature heteropterans have two pairs of wings and the forewings usually have the basal part hardened into a leathery texture; the membranous wing tip overlaps the tip of the other forewing over the abdomen.

The sub-order Auchenorrhyncha consists of about 350 British species, includes our one very rare Cicada and all the hopper bugs.

The sub-order Sternorrhyncha has over 750 species and includes the psyllids,

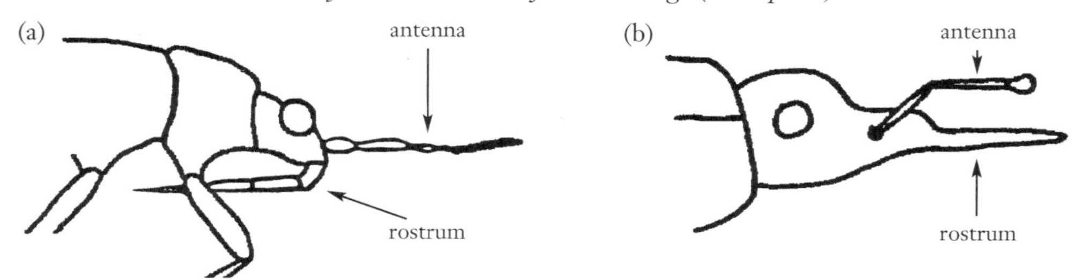

FIG.1 Bugs versus weevils
Bugs, (a), have the rostrum pointing to the rear, or downwards, and antennae that are neither clubbed nor 'elbowed' whilst weevils (b) have an unsegmented forward-pointing rostrum and clubbed, elbowed antennae.

whiteflies, aphids and scale insects. In the Auchenorrhyncha and Sternorrhyncha, both pairs of wings, if present, are generally membranous and are held roof-like over the abdomen when at rest. In older works, the Auchenorrhyncha and Sternorrhyncha will be found as 'series' under the sub-order Homoptera.

The insects with which bugs are most frequently confused are the weevils (Fig. 1b), which are beetles (Coleoptera, Curculionoidea). These have a forward-facing rostrum which is part of the head and not segmented. In a very few species, the rostrum is bent downwards but these have characteristic elbowed antennae which are expanded or clubbed at the tip. Like all beetles, they have the forewings completely hardened into elytra (wing cases) which meet on the centre-line of the abdomen without any overlap.

Cockroaches (Orthoptera) can bear a superficial resemblance to heteropteran bugs, but they may be distinguished by their slender multi-segmented antennae and biting mouthparts. Psocids or booklice (Psocoptera) can be mistaken for psyllids or aphids, but their antennae have at least 12 segments compared to the 10 of psyllids and up to 10 for the aphids (usually less). Psocids also have biting mouthparts, but confirming this character will take careful examination. Perhaps the most difficult group to distinguish from bugs are the small lacewings (Neuroptera) that are covered in a white waxy powder and look very much like whiteflies (Aleyrodidae). Lacewings do have biting mouthparts, but this is not an easy character to see. The commonest such lacewing, *Conwentzia psociformis* (Curtis), can be separated from the whiteflies by its much reduced hindwings. Whiteflies also have wings of equal size.

Collecting bugs

Bugs may be collected using butterfly or sweep nets, but sweep nets often seem more effective. The smaller bugs may also be sucked up in an aspirator or pooter. One of the simplest pooters is that designed by Disney (1999) which consists of two pieces of transparent plastic tubing of different sizes, one pushed into the end of the other and separated by a piece of fine netting to act as a filter. It should be pointed out that it is unwise to suck up insects from any decaying matter, although this situation is rarely met when collecting bugs. Bugs can also be obtained by static trapping, using such devices as Malaise, pitfall, suction or water traps. However, bugs do not spend as much time in flight as do flies (Diptera) and the Hymenoptera, so the numbers caught by trapping are usually relatively small. The simplest static trap is the water trap, which is simply a white or yellow bowl containing a little water to which a few drops of detergent have been added. Such a trap will collect some bugs, but one must be prepared to sort through a large number of other insects to find them.

It may be possible to identify some of the larger bugs using a hand lens and without killing them, but most bugs are small and it is necessary to kill them and use a microscope. A stereo dissecting microscope with a magnification of up to x25, or ideally x50, will be quite adequate for most purposes. The total magnification is found by multiplying the magnification of the objective lens system by that of the eyepiece.

The usual way of killing bugs is to use a killing jar. This can be made by casting a layer of plaster of Paris in the bottom of a jar or tube which will absorb a liquid killing agent, such as ethyl acetate. Some prefer to use a wad of cotton wool or tissue, or introduce a piece of absorbent material soaked in killing agent after the insect has been caught. Ethyl acetate is probably the best liquid killing agent because it leaves the animal more relaxed which helps when mounting, but there are alternatives such as ammonia or some household stain removers. Many volatile solvents will kill insects. To avoid degradation of the specimen, and in particular the wings, only the vapour of the killing agent should come into contact with the insect. Another method of killing bugs is to drop them into a tube of alcohol, and this will be convenient if storage in alcohol is intended, but some of the more delicate bugs should preferably be killed by exposure to the vapour of a liquid killing agent before being placed in alcohol to avoid damage.

Preserving bugs in spirit

Ecologists and others will generally wish to preserve their specimens in alcohol, and a 70% solution is usually found to be the best concentration. The addition of 5% of glycerol is a useful precaution against the specimen becoming too brittle in storage. The safest way of labelling spirit specimens is to write the information in hard pencil on a piece of thin card which is inserted into the tube. External labels have a habit of falling off.

Dry preservation: pinning and pointing

Those building up a collection will generally prefer some dry method of preservation. The easiest method is by pinning, and direct pinning may be used for the larger, more robust bugs. They should be pinned through the thorax, slightly off-centre, and the pin should hold the data label. Characters that may be obscured by the pin on one side may be seen on the other, but the central parts should not be damaged. For smaller bugs, indirect pinning (Fig. 2a) may be used. A small headless pin is used, again through the thorax and slightly off-centre, and the bug is pinned to some kind of support or stage. This may be a strip of plastic foam or even card, which is itself supported on a large entomological pin, which also holds the data label. An alternative to pinning is the technique known as pointing, where the bug is glued to a small triangular point of card by the side of the thorax (Fig. 2b). The card point is then held by a large entomological pin with its data label. Details of a suitable glue are given below.

Dry preservation: carding

All the above mounting techniques allow the underside of the bug to be examined easily, unlike the traditional method called carding, where the bug is glued down to a small rectangular card by its legs and antennae (Fig. 2c). Carding looks very tidy in a museum drawer, and it gives the specimen far greater protection against accidental damage than

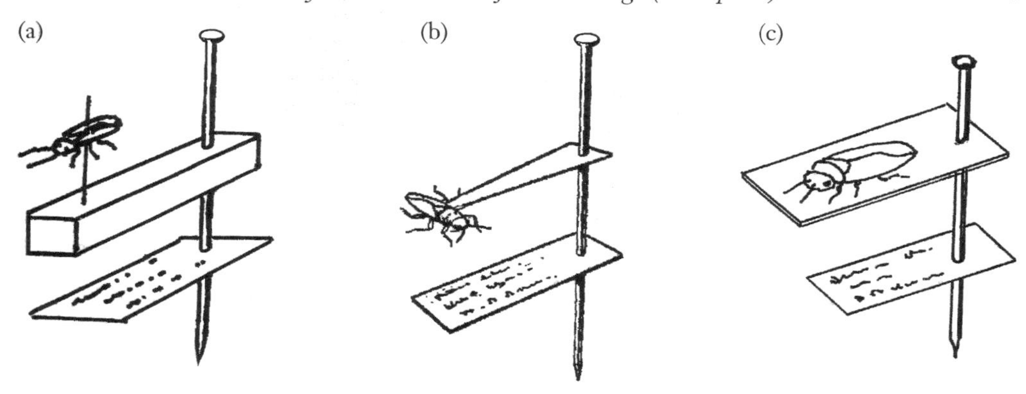

FIG. 2. Alternative techniques for dry mounting insects
(a) indirect pinning, (b) pointing and (c) carding.

pinning or pointing, but it makes the underside of the bug impossible to examine and the glue often obscures details of the legs and antennae. This may not matter to someone who is familiar with the group, but for others it will make identification difficult or impossible. If a thin water-based glue is used, then it can be dissolved and the insect removed for examination, but this carries the risk of accidental damage. If the bug is secured by a heavy, permanent glue such as Secotine, then removal may not be an option. If bugs are to be glued, then an absolute minimum of a water-based glue should be used. Such a glue can be made by dissolving a small quantity of Gum Tragacanth (available from larger chemists) in alcohol while stirring, until the mixture is about double the volume of the original powder. Then add four or five times this volume of water until a smooth paste is obtained. One gram of Gum Tragacanth will make up a small tube of glue that will fix a large number of bugs, and may also be used to repair specimens. When dry this glue will not obscure details in the tarsi (feet) or antennae of the bug. Some have used thin wallpaper paste for carding and although I have never tried it, the idea seems to be a good one since this is also water soluble.

Slide mounting

The small, soft-bodied bugs, such as aphids, psyllids and whiteflies, are not suitable for dry preservation since they generally shrivel when dried, and should generally be stored in alcohol or slide mounted. Details of slide mounting technique are given by Disney (1999).

IDENTIFICATION

This is a key to adult bugs. The nymphs (larvae or immature stages), of which there are usually five instars, are generally similar to the adults and have similar lifestyles, but do not have wings. However, some adult bugs have very poorly developed wings and it is not always easy to decide whether such a specimen is an adult or a nymph. In an adult the scutellum is a separate structure whereas, in a nymph, it is not clearly divided from the wing pads, as shown in Fig. 3. Many nymphs have characteristics sufficiently similar to the adults that they will key out correctly, but some will not. In particular, nymphs may have fewer antennal and tarsal segments.

There is a comprehensive and well-illustrated guide to the families of the Hemiptera (Dolling 1991), although it is unfortunately out of print at the time of writing. It includes

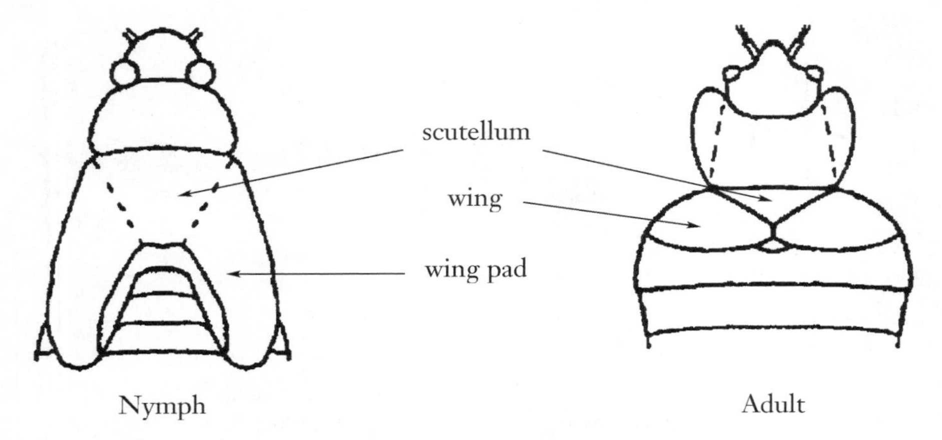

FIG. 3. The features used to separate nymphs from adults.

useful keys to the families of the aphids and scale insects, and the comprehensive bibliography is particularly helpful. This work incorporates a number of changes in classification and care should be taken when going on to older works where the family concepts may not be the same. To identify bugs beyond the family stage, a number of keys are available and those that are known about, at the time of writing, are give in the references on page 9.

The Royal Entomological Society of London publish a series of *Handbooks for the Identification of British Insects*, which may be obtained from the Registrar at 41 Queen's Gate, London SW7 5HT. They also publish a checklist (Kloet & Hincks, 1964) which is unfortunately out of print although a new checklist is in preparation at the time of writing. Another series of handbooks that include some very useful keys is the *Fauna Entomologica Scandinavica* series, details of which can be obtained from specialist book dealers.

The sub-order Heteroptera is covered by Southwood & Leston (1959), which is regrettably out of print. Although this book is quite old, it is really the only work available for the whole of this group and is well illustrated by colour plates. Fortunately, it was a popular book and there are many copies in libraries.

In the Sternorrhyncha, the Aphidoidea is a difficult but important group and the most useful starting point is a little book by Blackman (1974). More comprehensive accounts are given by Blackman & Eastop (1985) and Blackman & Eastop (1994). There are several relevant volumes in the *Fauna Entomologica Scandinavica* and *Handbooks for the Identification of British Insects* series.

A note on scale insects (Coccoidea)

The scale insects are very strange animals. They don't look much like bugs; they are very diverse, and many don't look like insects at all! The females are always wingless, their bodies are hardly segmented, they have simple eyes (ocelli) rather than the usual insect compound eye, and some are devoid of legs and antennae. Adult males are usually about 1mm long and do at least look like insects since they have wings, but they have only one pair and the hindwings are reduced to short stalks that hook onto the front wings; they are often assumed to be tiny two-winged flies (order Diptera). They often have compound eyes, but they have no mouthparts, so they certainly don't look like bugs (Fig. 4).

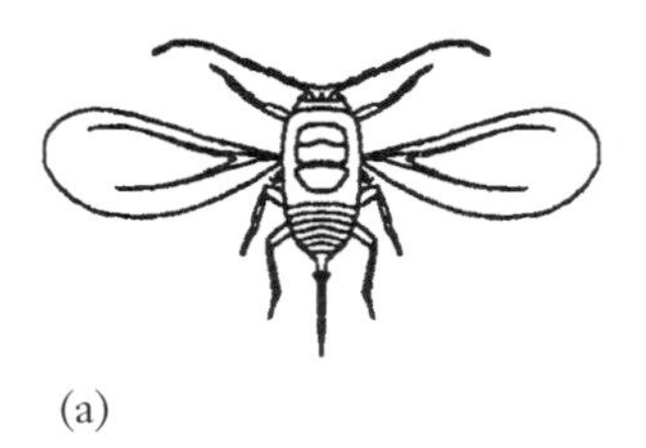

(a)

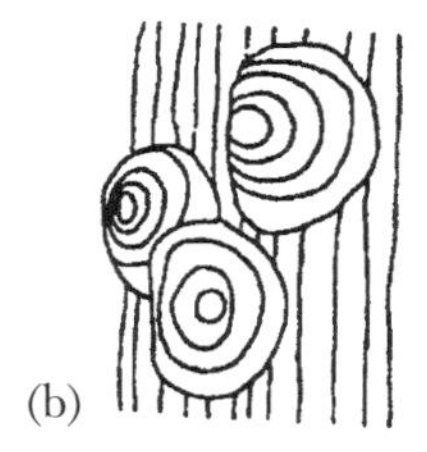

(b)

(c)

FIG 4. Scale insects
(a) a male of the family Diaspididae, (b) three females on a plant stem and (c) an attractive female of the family Ortheziidae

Female scale insects are just as extraordinary as the males. Many protect themselves with a hard waxy scale or a mass of waxy threads, and it is this waxy scale that gives the group its common name. The scales vary enormously in size, shape and colour, and it is by these that much of the identification is done. The females shown in Fig. 4b and c are just two of a huge range of scale insects and are in no way representative of the group as a whole. There are even some that do not protect themselves with a scale, including the mealybugs (Pseudococcidae). These are distinctly segmented, covered in mealy wax, and have fully developed legs.

There are nine families of scale insects (Coccoidea) shown in Kloet & Hincks (1964) although older authors recognise only one, the Coccidae. What were once sub-families have been raised to family rank and there have been many changes in nomenclature. The best starting point is Dolling (1991) and help may also be obtained from Greene (1922). There is a Ray Society monograph by Newstead (1901-1903) but, although this old work contains much useful information, careful use of the checklist will be needed because of the many new species and name changes since 1901. Such characters as the scale insects do possess are mainly visible only with a high powered microscope and using prepared and slide mounted specimens. Such identification is outside the scope of this key.

Scale insects are prolific breeders and can become serious horticultural pests. They may be difficult to deal with simply because the scales give the females so much protection from insecticide sprays. In some cases, scale insects have been imported from other continents with their host plants, and, since we do not have appropriate predators, their populations can grow at alarming rates, particularly in greenhouses.

As far as this key is concerned, scale insects look so little like other bugs, or so little like other insects, that one cannot know that they are bugs at all unless one already knows that they are scale insects. If you are sure that you have a scale insect, there is little point in putting it through the key, but for the sake of completeness they have been included.

ACKNOWLEDGEMENTS

I am indebted to Professor Michael Akam for the use of the insect collection and library at the Cambridge University Museum of Zoology, and to the curator of insects, Dr William Foster for his help and encouragement. I would also like to thank Dr Sarah Corbet and Dr Henry Disney for their comments on the manuscript. I particularly acknowledge my debt to all those kind people who have tested this key and made so many helpful comments. Finally my thanks to Dr Steve Tilling and Dr Rebecca Farley for guiding the key through the process of testing and publication.

TABLE OF SUB-ORDERS, SUPER-FAMILIES AND FAMILIES RECOGNISED IN THIS KEY

Names follow Dolling (1991).

HEMIPTERA

Heteroptera

Acanthosomatidae
Aepophilidae
Alydidae
Anthocoridae
Aphelocheiridae
Aradidae
Berytidae
Ceratocombidae
Cimicidae
Coreidae
Corixidae
Cydnidae
Dipsocoridae
Gerridae
Hebridae
Hydrometridae
Lygaeidae
Mesoveliidae
Microphysidae
Miridae
Nabidae
Naucoridae
Notonectidae
Nepidae
Pentatomidae
Piesmidae
Pleidae
Pyrrhocoridae
Reduviidae
Rhopalidae
Saldidae
Scutelleridae
Stenocephalidae
Tingidae
Veliidae

Auchenorrhyncha

Cercopidae
Cicadellidae
Cicadidae
Cixiidae
Delphacidae
Issidae

Membracidae
Tettigometridae

Sternorrhyncha

Adelgoidea

Adelgidae
Phylloxeridae

Aleyrodoidea

Aleyrodidae

Aphidoidea

Anoeciidae
Aphididae
Callaphididae
Chaitophoridae
Hormaphididae
Lachnidae
Mindaridae
Pemphigidae
Phloeomyzidae
Thelaxidae

Coccoidea

Psylloidea

Calophyidae
Homotomidae
Liviidae
Psyllidae
Spondyliaspidae
Triozidae

Further Reading and References

BLACKMAN, R., (1974). *Aphids*. Ginn and Company Ltd., London and Aylesbury.

BLACKMAN, R. L. & EASTOP, V. F., (1985). *Aphids on the world's crops*. John Wiley and Sons, New York and London.

BLACKMAN, R. L. & EASTOP, V. F., (1994). *Aphids on the world's trees*. CAB International, Oxford.

CHINNERY, M., (1973). *A field guide to the Insects of Britain and Northern Europe*. William Collins, London.

DISNEY, R. H. L., (1999). *British Dixidae (Meniscus Midges) and Thaumaleidae (Trickle Midges): Keys with Ecological Notes*. Freshwater Biological Association Scientific Publication No. 56, 129pp.

DOLLING, W. R., (1991). *The Hemiptera*. Oxford University Press, Oxford.

DOUGLAS, J. W. & SCOT, J., (1865). *The British Hemiptera Vol. 1 Hemiptera-Heteroptera*. Ray Society, London.

EDWARDS, H. J., (1896). *Hemiptera Homoptera*. L. Reeve and Co., London.

GIBBONS, Bob, (1995). *Field guide to the insects of Britain and Northern Europe*. Crowood Press, Marlborough, Wiltshire.

GREENE, E. E., (1922). A brief review of the indigenous Coccidae of the British Islands. *South London Entomological and Natural History Society. Proceedings*. pp 12–25.

HEIE, O. E., (1980). *The Aphidoidea (Hemiptera) of Fennoscandia and Denmark. I. General part. The families Mindaridae, Hormaphididae, Thelaxidae, Anoeciidae, and Pemphigidae*. Fauna Entomologica Scandinavica Vol. 9.

HEIE, O. E., (1982). *The Aphidoidea (Hemiptera) of Fennoscandia and Denmark. II. The family Drepanosiphidae.* Fauna Entomologica Scandinavica Vol. 11.

HEIE, O. E., (1986). *The Aphidoidea (Hemiptera) of Fennoscandia and Denmark. III. Family Aphididae: subfamily Pterocommatinae and Tribe Aphidini of subfamily Aphidinae.* Fauna Entomologica Scandinavica Vol. 17.

HEIE, O. E., (1991). *The Aphidoidea (Hemiptera) of Fennoscandia and Denmark. IV. Family Aphididae: Part 1 of Tribe Macrosiphini of subfamily Aphidinae.* Fauna Entomologica Scandinavica Vol. 25.

HEIE, O. E., (1994). *The Aphidoidea (Hemiptera) of Fennoscandia and Denmark. V. Family Aphididae: Part 2 of Tribe Macrosiphini of subfamily Aphidinae.* Fauna Entomologica Scandinavica Vol. 28.

HEIE, O. E., (1995). *The Aphidoidea (Hemiptera) of Fennoscandia and Denmark. VI. Family Aphididae: Part 3 of Tribe Macrosiphini of subfamily Aphidinae, and the family Lachnidae.* Fauna Entomologica Scandinavica Vol. 31.

HODKINSON, I. D. & WHITE, I. M., (1979). *Homoptera, Psylloidea.* Handbooks for the Identification of British Insects, Vol. 2, part 5a. Royal Entomological Society of London.

JERVIS, M., STEWART, A., & WILSON, A., (in prep.) *Leafhoppers.* Naturalists' Handbooks. Richmond Publishing Co. Ltd., Slough.

KIRBY, P., (1992). *A review of the scarce and threatened Hemiptera of Great Britain.* Research and Survey in Nature Conservation Series. Nature Conservancy Council, Peterborough.

KLOET, G. S. & HINCKS, W. D., (1964). *A checklist of British Insects. Small orders and Hemiptera.* Handbooks for the Identification of British Insects, Vol. 11, part 1. Royal Entomological Society of London.

LE QUESNE, W. J., (1960). *Hemiptera (Fulgoromorpha).* Handbooks for the Identification of British Insects, Vol. 2, part 3. Royal Entomological Society of London. (This covers the families Cixiidae, Delphacidae, Issidae and Tettigometridae.)

LE QUESNE, W. J., (1965). *Hemiptera, Cicadomorpha (exc. Deltocephalinae and Typhocybinae).* Handbooks for the Identification of British Insects, Vol. 2, part 2a. Royal Entomological Society of London. (This covers the families Cicadidae, Cercopidae, and parts of Cicadellidae.)

LE QUESNE, W. J., (1969). *Hemiptera (Cicadomorpha – Deltocephalinae).* Handbooks for the Identification of British Insects, Vol. 2, part 2b. Royal Entomological Society of London.

LE QUESNE, W. J. & PAYNE, K. R., (1981). *Cicadellidae (Typhocybinae) with a checklist of the British Auchenorrhyncha (Hemiptera, Homoptera).* Handbooks for the Identification of British Insects, Vol. 2, part 2c. Royal Entomological Society of London.

NEWSTEAD, R., (1901–1903). *Monograph of British Coccidae.* (2 volumes). Ray Society, London.

OSSIANNILSSON, F., (1978). *The Auchenorrhyncha (Homoptera) of Fennoscandia and Denmark. Part 1: Introduction, infraorder Fulgoromorpha.* Fauna Entomologica Scandinavica Vol. 7, part 1.

OSSIANNILSSON, F., (1981). *The Auchenorrhyncha (Homoptera) of Fennoscandia and Denmark. Part 2: The families Cicadidae, Cercopidae, and Cicadellidae (excl. Deltocephalinae).* Fauna Entomologica Scandinavica Vol. 7, part 2.

OSSIANNILSSON, F., (1983). *The Auchenorrhyncha (Homoptera) of Fennoscandia and Denmark. Part 3: The Family Cicadellidae: Deltocephalinae, Catalogue, Literature and Index.* Fauna Entomologica Scandinavica Vol. 7, part 3.

SAUNDERS, E., (1892). *The Hemiptera-Heteroptera of the British Islands.* L. Reeve and Co., London.

SAVAGE, A. A., (1989). *Adults of British aquatic Hemiptera Heteroptera: a key with ecological notes.* Freshwater Biological Association Scientific Publications No. 50.

SOUTHWOOD, T. R. E. & LESTON, D., (1959). *Land and water bugs of the British Isles.* Frederick Warne & Co. Ltd., London.

STROYAN, H. L. G., (1977). *Aphidoidea – Chaitophoridae and Callaphididae.* Handbooks for the Identification of British Insects, Vol. 2, part 4a. Royal Entomological Society of London.

STROYAN, H. L. G., (1984). *Aphids – Ptercommatidae and Aphidinae (Aphidini). (Homoptera, Aphidoidea).* Handbooks for the Identification of British Insects, Vol. 2, part 6. Royal Entomological Society of London.

TILLING, S. M., (1987). A key to the major groups of British terrestrial invertebrates. *Field Studies*, **6**, 695–766.

TERMS USED TO DESCRIBE PARTS OF A BUG

The terms used to describe the more important parts of a bug are given in the figure below. The abdomen is hidden by the wings in this view. Other terms will be explained at appropriate points in the key.

NOTE that the segments of the tarsi (feet), rostrum (beak) and antennae are numbered from the end nearest the body and, when counting tarsal segments, the claws are not counted as a separate segment.

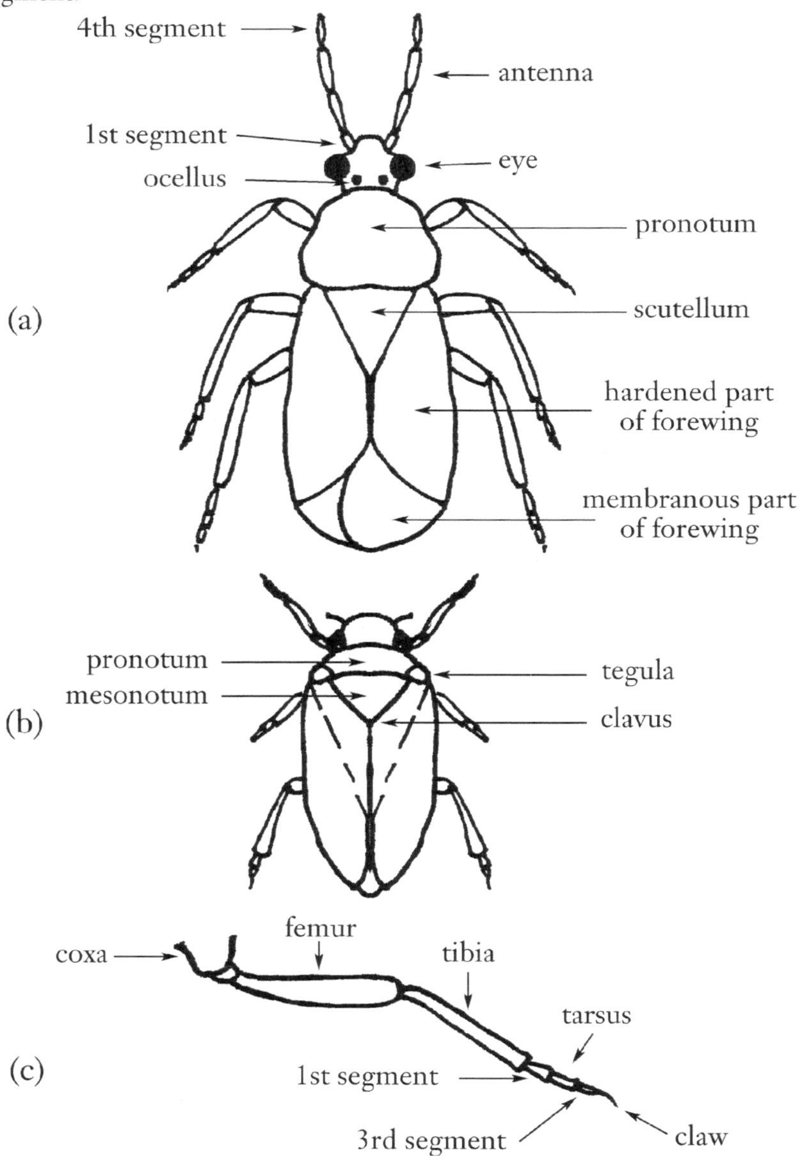

FIG. 5. The terms used to describe the parts of a bug used in this key.
(a) Sub-Order Heteroptera (b) Sub-Order Auchenorryncha (c) a leg

KEY TO SUB-ORDERS WITHIN THE HEMIPTERA

The order Hemiptera now has three sub-orders; the Heteroptera, the Auchenorrhyncha and the Sternorrhyncha. In older taxonomic works, only two sub-orders were recognised, the Heteroptera and the Homoptera, but the latter has now been split by elevating the series Auchenorrhyncha and the series Sternorrhyncha to the rank of sub-order. For reasons mentioned earlier, it is not intended that the scale insects should be put through this key but, in case they are not recognised, a note has been placed at the point where they will emerge.

1 Antennae inserted underneath the head (ventrally), and much shorter than the head; generally invisible from above...............Sub-order **Heteroptera** (part)(page 14)
Aquatic bugs living below the water surface.

\- Antennae inserted in the side or front of the head, usually longer than the head and normally visible from above. **2**

2(1) Antennae consisting of a few short segments with a terminal unsegmented arista (bristle) that is much longer than any of the segments (Fig. 6). Head extends backwards underneath (postero-ventrally) so that the rostrum (beak) arises from between the front legs.

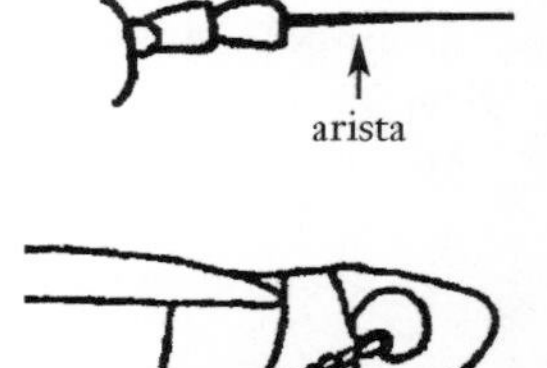

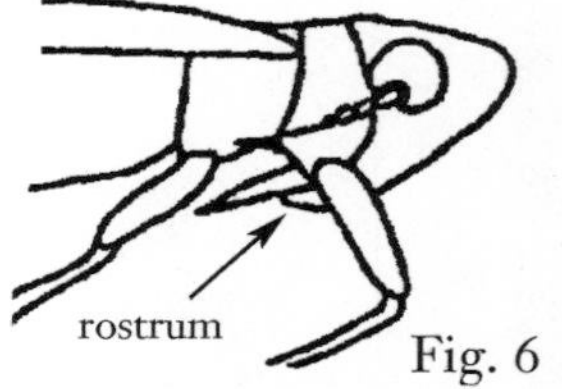

Fig. 6

.... Sub-order **Auchenorrhyncha** (page 26)

\- Antennae normally with 4 or 5 similar segments (Fig. 7a), (occasionally up to 10) but never with an unsegmented terminal arista as illustrated above. [Some aphids have a terminal process on the last segment, not a tapering bristle, but these antennae have long slender segments throughout (Fig. 7b). Some psyllids have the basal segments expanded, but the rest of the antenna is segmented.] Rostrum arising from the front or underside of the head, but always ahead of the front legs (Fig. 7).

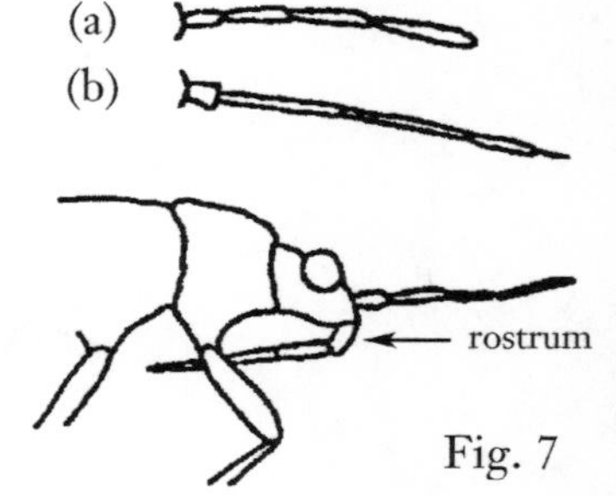

Fig. 7

.... **3**

3(2) Rostrum present and arising from the front of the head (Fig. 8). Wings generally present and divided into two distinct regions. The basal two-thirds is hardened and leathery, while the wingtip is clear and membranous. The wings are folded flat over the body when at rest, and the clear wingtip area overlaps the other forewing. There are a very few species which have the whole wing completely hardened or with the wing mostly clear

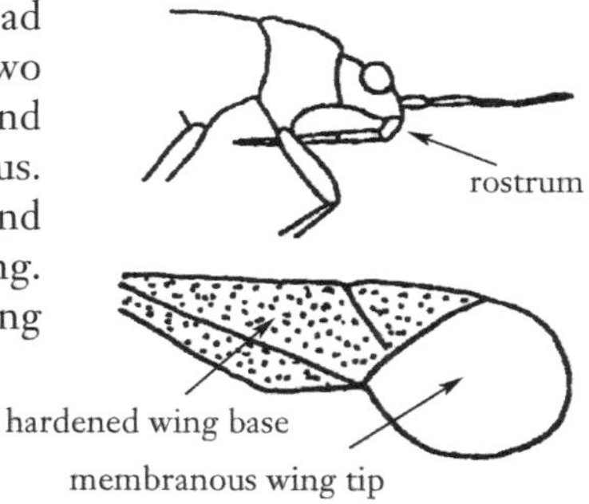

Fig. 8

.... Sub-order **Heteroptera** (opposite)

A large and diverse group of insects, many of which are somewhat flattened, which includes all the true water bugs, the predatory and bloodsucking bugs, and the plant bugs.

\- Rostrum is occasionally absent, but when present arises from below the head but ahead of the front legs. When wings are present, the forewings are of the same texture throughout, and are usually held roof-like over the body when at rest. If they are held flat, then the tip of one forewing does not overlap the other forewing.

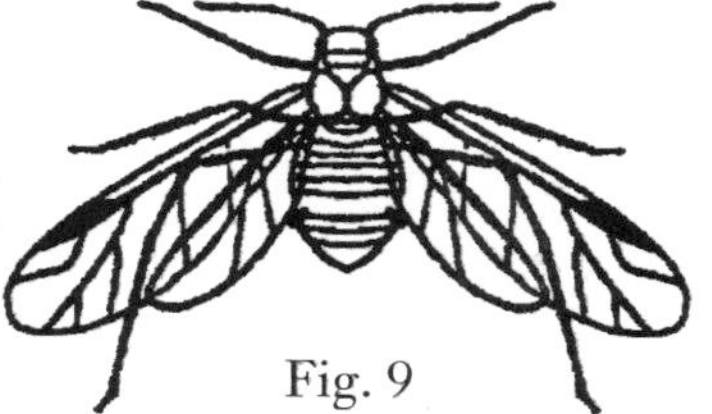

Fig. 9

.... Sub-order **Sternorrhyncha** (page 29)

Bugs in this group are generally small or minute, and are all plant feeders. They include greenflies and blackflies (aphids) (Figs 9, 10a), whiteflies (Fig. 10b), psyllids (Fig. 10c), and scale insects. Note that the left hindwing of the whitefly illustrated in Fig. 10b is hidden below the forewing.

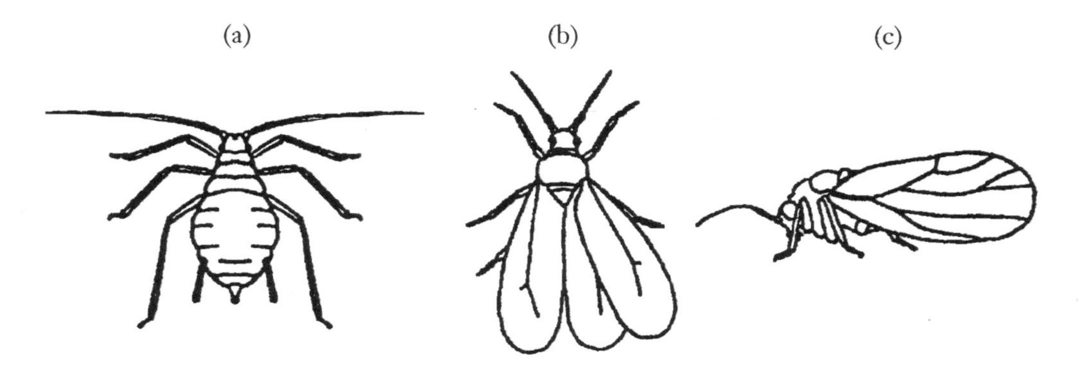

Fig. 10

KEY TO FAMILIES IN THE HETEROPTERA

1 Antennae inserted underneath the head (ventrally), and much shorter than the head, generally invisible from above. True water bugs living below the surface.. **2**

\- Antennae inserted at the front or side of the head, and longer than the head, clearly visible from above. Land and amphibious bugs, including those living on the water surface. **7**

2(1) Bugs with a long, stiff tail (Fig. 11). (NOTE that the 'tail' consists of two components that combine to form a respiratory siphon. These may become detached in dead specimens, giving the appearance of two tails.) Front legs modified to seize prey (raptorial). .... **Nepidae**

Two British species:

Nepa cineria L., (Fig. 11a) the WATER SCORPION, is a large predatory bug that is found in shallow pools, ditches, and the margins of slow moving rivers, and feeds on insects, tadpoles and small fish. 18–22mm long.

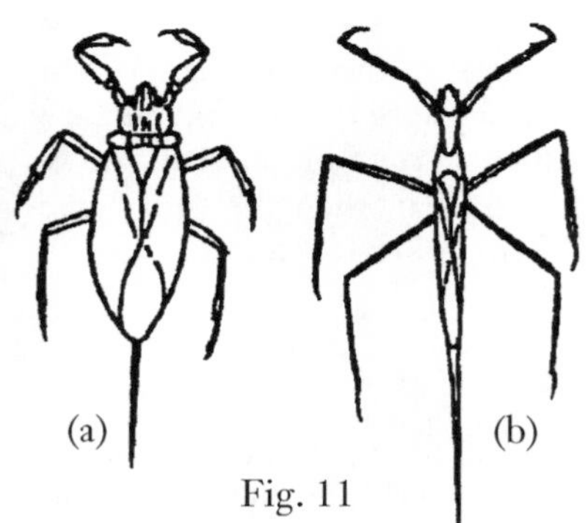

Fig. 11

Ranatra linearis (L.), (Fig. 11b), the WATER STICK INSECT, is an even longer bug, found in deep pools, water-filled quarries, and slow rivers among emergent vegetation, and feeds on small arthropods. 30–35mm long.

\- Abdomen without such a tail. **3**

3(2) Each pair of legs modified in some way, and structurally different to the others. Tarsi of middle pair with one segment and very long claws (Fig. 12). Front tarsi with one segment. Rostrum very short and broad. .... **Corixidae**

Fresh water bugs, often known as LESSER WATER BOATMEN, that mostly feed on plant debris and algae, but some are predatory. 8 genera, 33 species, 3–13mm long.

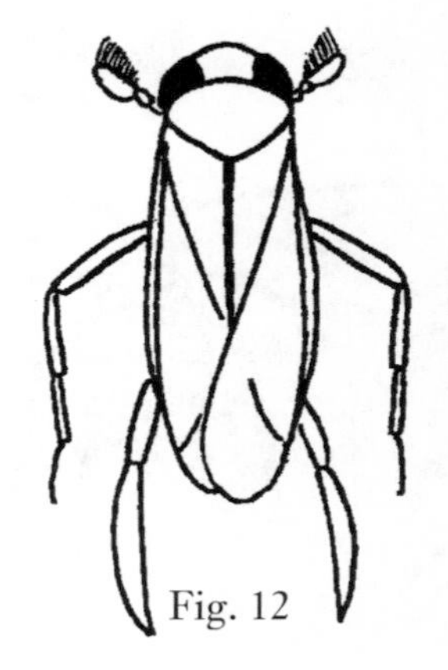
Fig. 12

\- With at most one pair of legs much modified in comparison with the others (see couplets 4–6 below). Rostrum pointed. **4**

4(3) Front femora greatly swollen with tibiae folded against them (Fig. 13). Outer margin of eye forming a continuous curve with the front of the head and the lateral margin of the pronotum (Fig 13) .... **Naucoridae**

The SAUCER BUG. 1 species, *Ilyocoris cimicoides* (L.), 10–15mm. Middle and hind tibiae cylindrical with rows of bristles and weak swimming hairs. Found in ditches, slowish river margins and in gravel pits.

- Front femora not swollen as in Fig. 13. Eyes more prominent.. **5**

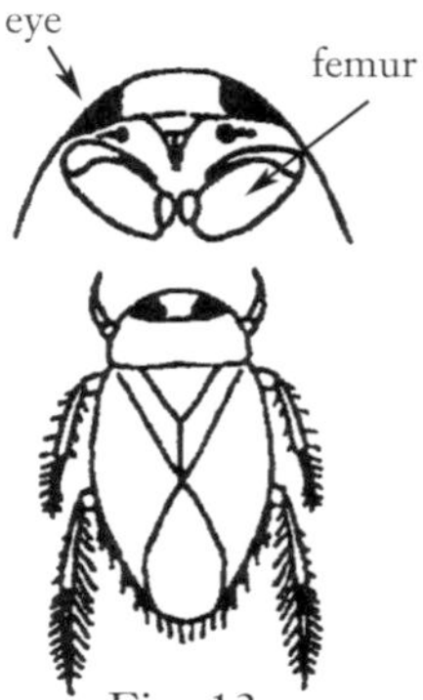

Fig. 13

5(4) Flattened, oval bug 8–10mm long with the forewings not fully developed (in the UK) (Fig. 14). Head as broad as long .... **Aphelocheiridae**

1 species, *Aphelocheirus aestovalis* (Fab.). Found in stony or gravely rivers and feeds on insect larvae, particularly midges.

- Large (over 10mm) boat-shaped bugs, or small (less than 3mm) domed bugs. Wings fully developed. Head broader than long. **6**

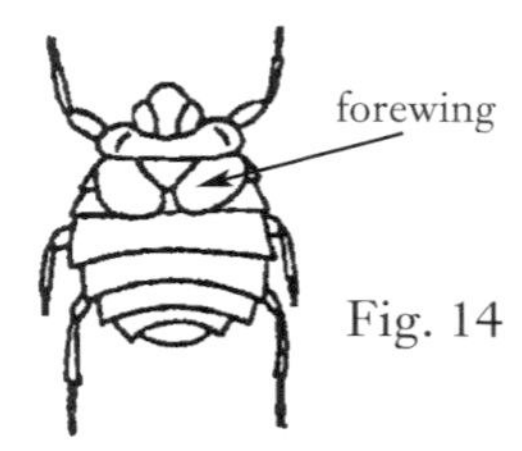

Fig. 14

6(5) Length 11mm or more. Boat-shaped bugs that swim on their backs. Posterior tibiae flattened and tarsi with two segments. Pronotum and wings smooth. .... **Notonectidae**

WATER BOATMEN. 1 genus, *Notonecta* L., (Fig. 15) and 4 species. Predatory bugs inhabiting slow moving or still water.

Fig. 15

- Small bugs, not exceeding 3mm. Body strongly convex. Posterior tibiae cylindrical and tarsi with three segments. Pronotum and forewings (Fig. 16) covered in coarse pits .... **Pleidae**

LESSER WATER BOATMAN. 1 species, *Plea atomaria* (Pallas), found in pools, canals and slow rivers.

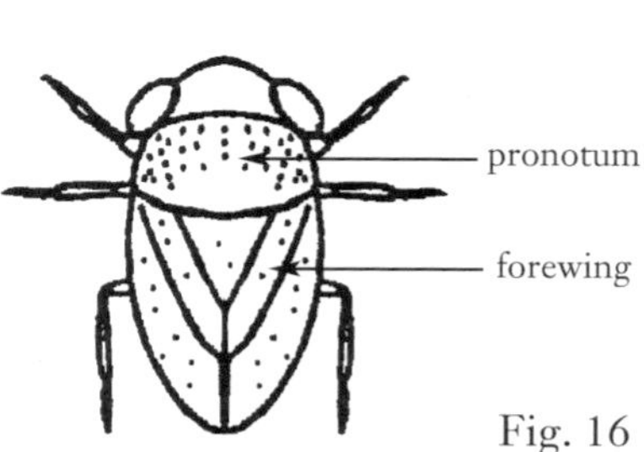

Fig. 16

7(1) Head much longer than wide. Very slender elongated bugs. (Fig. 17).. **Hydrometridae**

The WATER MEASURER. 1 genus, *Hydrometra* Latreille, 2 species. 7–12mm. Found on the water surface at the margins of streams, or on still or slow moving water. It spears its prey through the surface film.

- Head not more than twice as long as wide. **8**

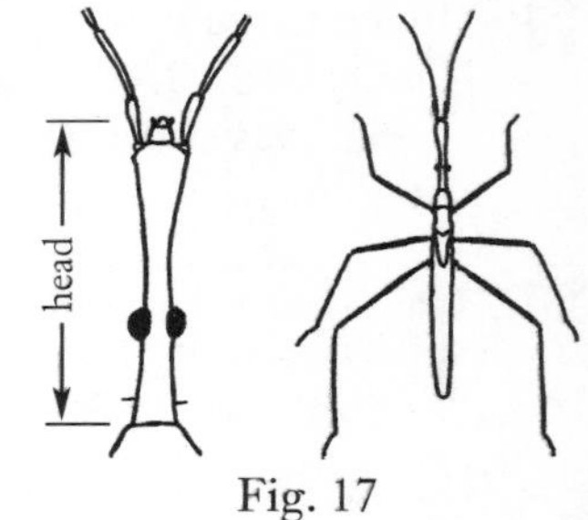

Fig. 17

8(7) Antennae with 5 segments (Fig. 18a), or in the case of bugs not exceeding 2mm in length, with the fourth segment constricted in the middle giving the appearance of five segments (Fig. 18b)........................... **9**

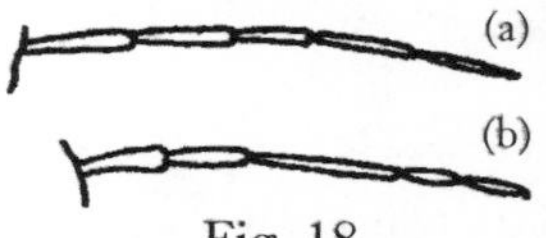

Fig. 18

- Antennae with 4 segments, the fourth segment never constricted in the middle (Fig. 19)............................ **15**

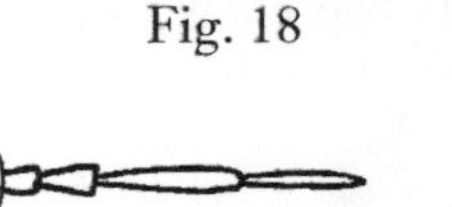

Fig. 19

9(8) Bugs not exceeding 2mm in length. Underside of abdomen covered with dense silvery hairs......... **Hebridae**

The SPHAGNUM BUG, 1 genus, *Hebrus* Curtis (Fig. 20), 2 species. Found in streams and torrents, usually in *Sphagnum* moss.

- 3mm or longer, underside of abdomen without a dense covering of hairs ....SHIELD BUGS.... **10**

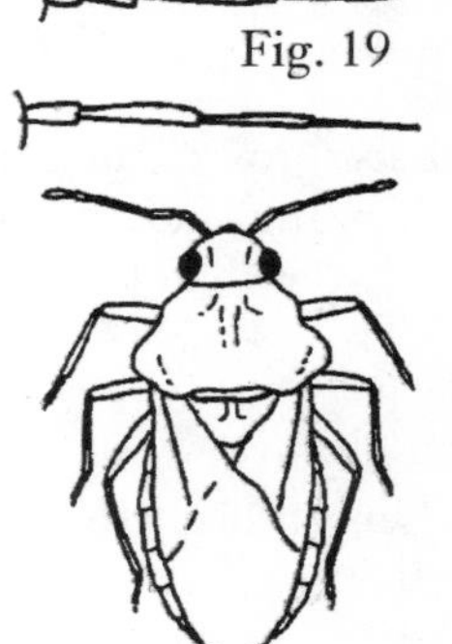

Fig. 20

10(9) Tarsi (feet) with 2 segments. Front end of the underside of the abdomen with a long, forward-facing spine (Fig. 21), extending almost to the point of insertion of the front legs...**Acanthosomatidae**

This family includes the hawthorn shieldbug. 3 genera, 5 species. 8 – 15mm long.

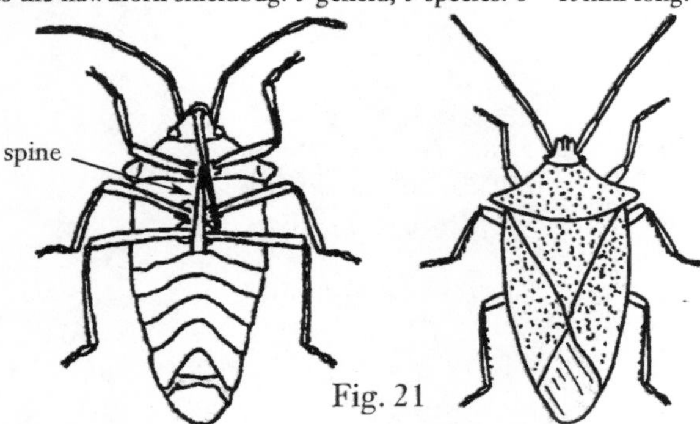

Fig. 21

- Tarsi with 3 segments. Front part of the underside of the abdomen without such a spine, or with a shorter one that does not reach beyond the point of insertion of the middle legs. **11**

11(10) Tibiae with strong spines (Fig. 22) **12**

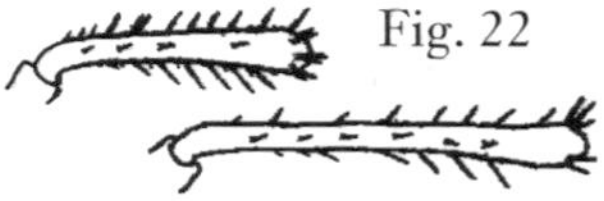

- Tibiae without strong spines. **13**

Fig. 22

12(11) Bugs never hairy. Black or dark, usually metallic bugs, sometimes with pale markings.................. **Cydnidae**
5 genera, 8 species. 3–9mm long. Feed on roots of low growing plants.

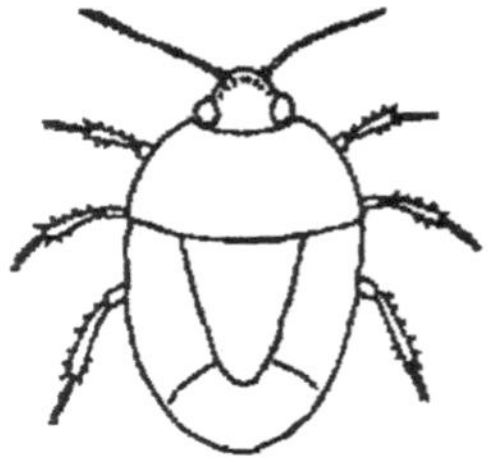

- Conspicuously hairy brown bugs. **13**

Fig. 23

13(11 or 12) Scutellum triangular, not extending to the end of the abdomen (Fig. 24). .... **Pentatomidae**
17 genera, 20 species. 4–14mm long. (part)

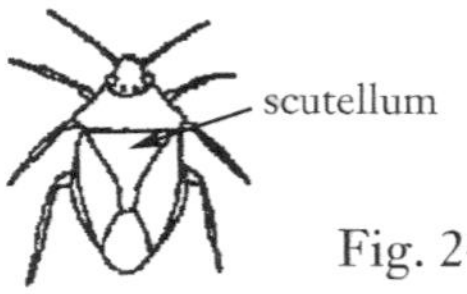

- Scutellum extending to the end of the abdomen (Fig. 25) **14**

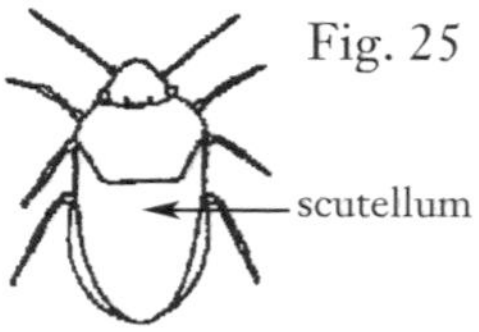

scutellum

Fig. 24

Fig. 25

scutellum

14(13) Margin of pronotum with a pair of curved projections beside the head (Fig. 26). 5–7mm long .... **Pentatomidae**
One species, *Podops inuncta* (Fab) will key out here. (part)

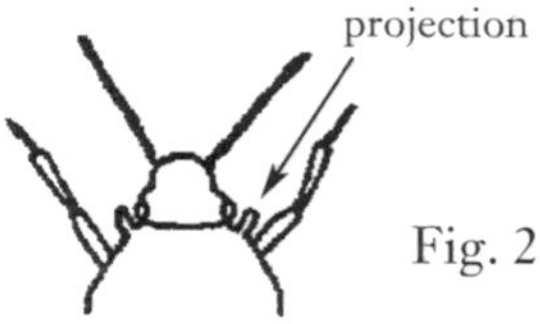

projection

Fig. 26

- Margin of pronotum without such projections (Fig. 27) 4–11mm long .... **Scutelleridae**
2 genera, 5 species.

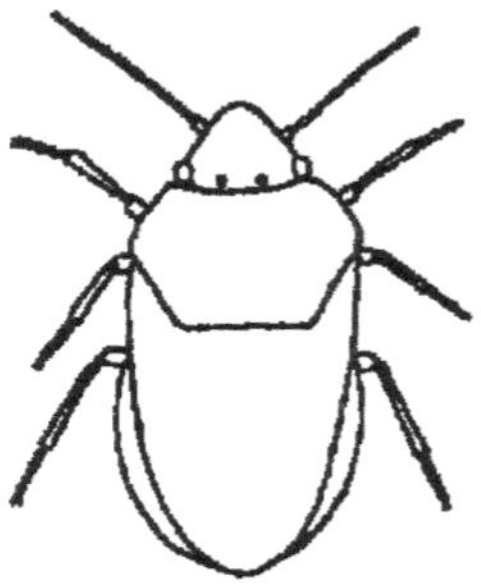

Fig. 27

15(9) Much of the underside of the abdomen covered with short fine hairs (silvery pubescence) giving a frosted appearance. Claws apical or subapical. Bugs associated with the water surface....... **16**

- Underside without such silvery pubescence. Claws always inserted at the tip and never before the tip (subapical) (see Fig. 28). Various habitats including water **18**

16(15) Claws of tarsi inserted at the tip (apical) (Fig. 28). .... **Mesoveliidae**

The POND WEED BUG, 1 species, *Mesovelia furcata* Mulsant & Rey (Fig. 29) 3–4mm long. Wings often reduced or absent. Olive-green bugs with dark markings. Found on the floating leaves of pondweed and feeds on insects on the surface.

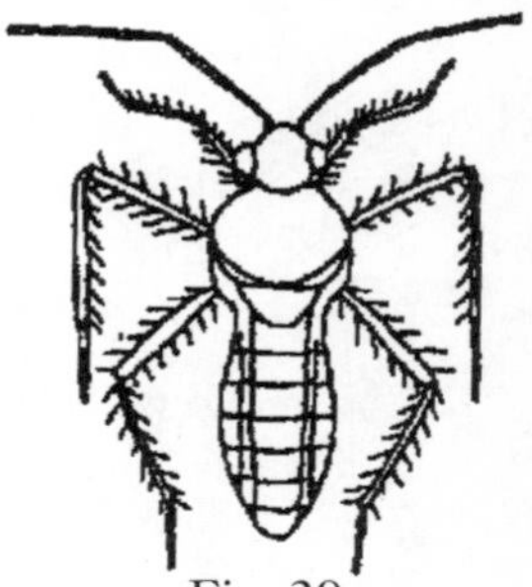

Fig. 28

Fig. 29

- Claws inserted before the actual tip (sub-apical) (Fig. 30) **17**

Fig. 30

17(16) Hind femora extending to beyond the tip of the abdomen (Fig. 31). Middle legs inserted close to hind legs, and well separated from front legs............ **Gerridae**

POND SKATERS. 3 genera, 10 species. 6–18mm long. Predators living on the water surface.

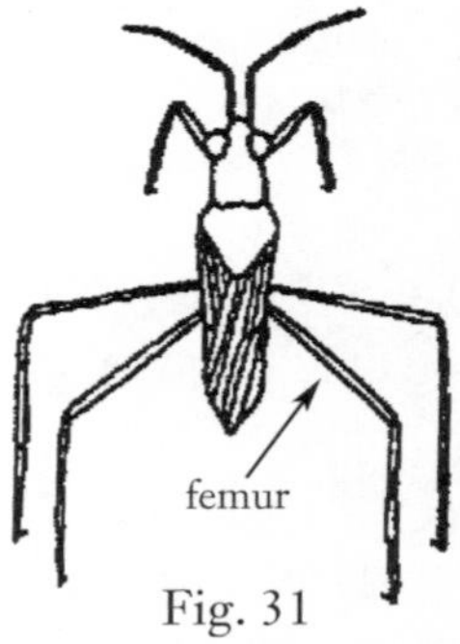

femur

Fig. 31

- Hind femora shorter, not extending beyond the abdomen (Fig. 32). Middle legs inserted about mid-way between front and hind legs .... **Veliidae**

WATER CRICKETS. 2 genera, 5 species. 2 – 7mm long. Found on the surface of ponds and streams, where they feed on arthropods that fall onto the water surface.

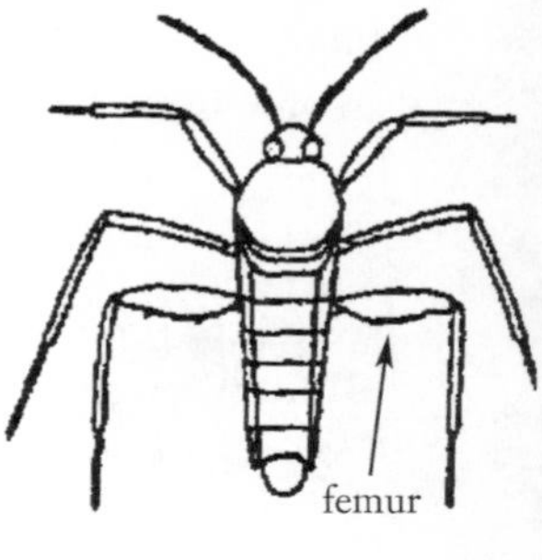

femur

Fig. 32

18(15) Tarsi (feet) with 2 segments (Fig. 33). Remember that claws are not segments **19**

Fig. 33

- Tarsi with 3 segments (Fig. 34) **22**

Fig. 34

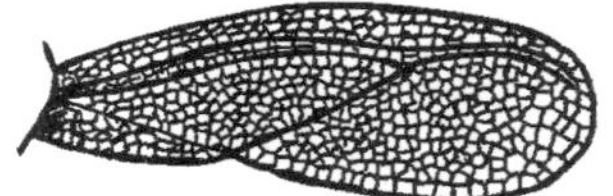

Fig. 35

19(18) Forewings fully developed and covered by a network of raised veins enclosing glassy areas that extends to the wingtip (Fig. 35). Pronotum covered with a similar pattern (Fig. 36). Ocelli absent ....**Tingidae**
LACE BUGS, found on a variety of plants including gorse, thistles, rhododendron, bugle and hawthorn. Some live in moss. 13 genera, 24 species. 2–5mm long.

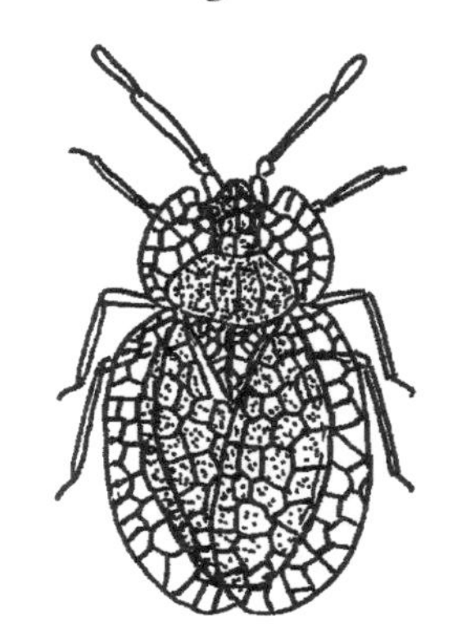

Fig. 36

- Forewings, if fully developed, with the tip always membranous (Fig. 37), but the hardened part of the wing may be covered in a dense pattern of punctures (Fig. 38) **20**

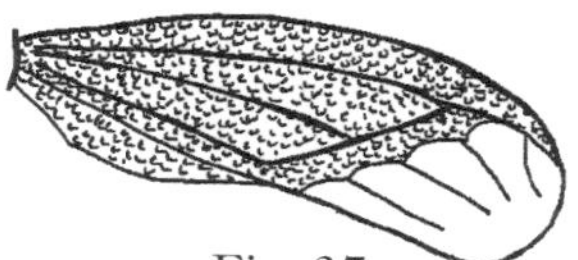

Fig. 37

20(19) Hardened part of forewing covered in a dense pattern of punctures but with the wingtip (Fig. 37) membranous, with veins. Pronotum covered with a similar pattern (Fig. 38). Ocelli present
.... **Piesmidae**
1 genus, *Piesma* le Peletier & Serville, 2 species. 2–3mm long. Found in coastal areas, gravel pits, arable land and waste sites. (Given as Piesmatidae in Kloet & Hincks, 1964.)

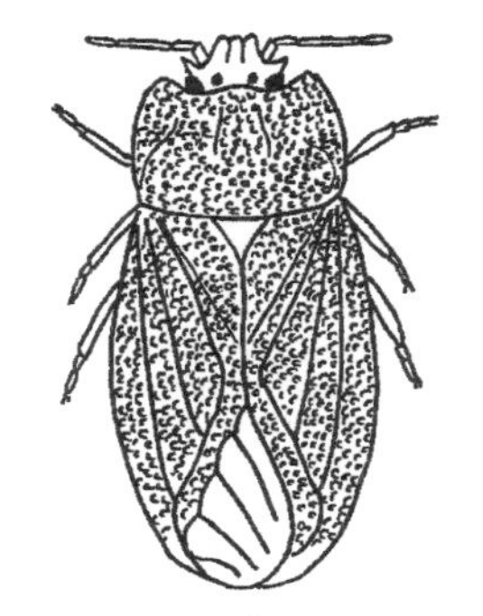

Fig. 38

- Forewings, if present, and pronotum without a dense pattern of punctures. **21**

21(20) Bugs over 3.5mm long. Very flattened, leathery bugs with short, robust antennae and conspicuously short legs. Living under tree bark and feeding mostly on fungi.................................... **Aradidae**

FLATBUGS. 2 genera, 7 species. 3.5–9mm long. Most feed on fungi, but the pine flatbug feeds on plant sap.

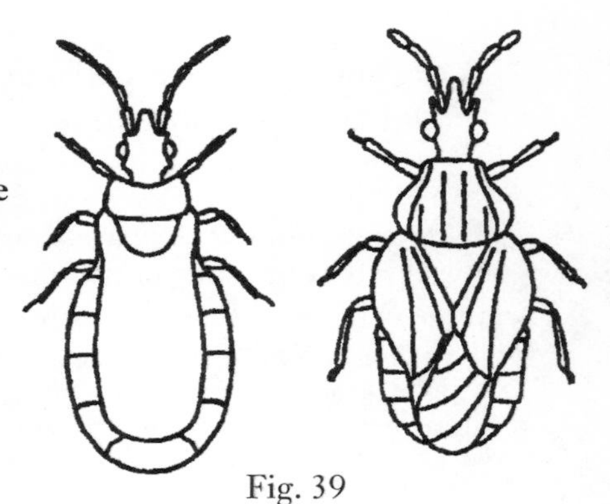

Fig. 39

- Bugs 2mm or less in length. Delicate bugs with long, thin antennae and long legs (Fig. 40). Often found on tree trunks and believed to feed on tiny arthropods. .... **Microphysidae**

2 genera, 6 species. Feed on small arthropods such as aphids, psocids, and springtails.

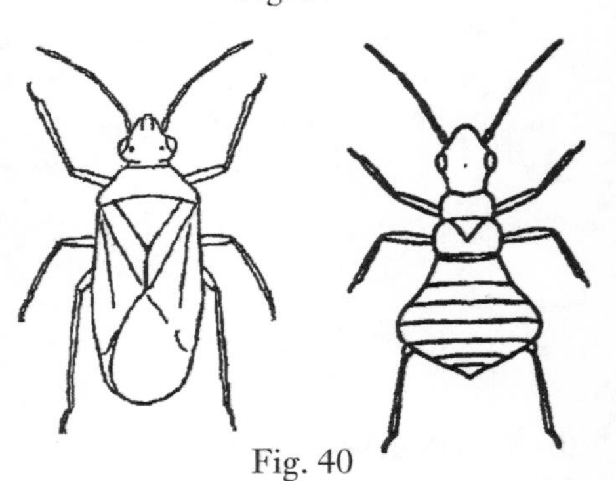

Fig. 40

22(18) Antennae (Fig. 41a) with segments 1 and 2 short and thick, segments 3 and 4 long and slender. Predatory bugs under 3mm long. .... **Dipsocoridae** and **Ceratocombidae**

Two families found in damp places (Fig. 41b). Dipsocoridae (2 genera, 2 species) have a short rostrum that does not extend beyond the prosternum (first segment of the thorax). Ceratocombidae (1 species, *Ceratocombus coleoptratus* (Zett)) have a longer rostrum extending almost to the point of insertion of the middle legs.

- Antennal segments more uniform, or bugs at least 3mm long. **23**

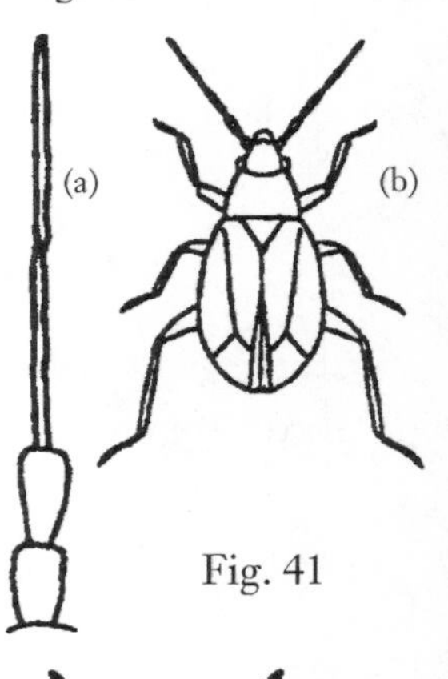

Fig. 41

23(22) Bugs about 3mm long, with a characteristic concave hind margin to the very short forewings (Fig. 42). .... **Aepophilidae**

1 species, *Aepophilus bonnairei* Signoret. Reddish-brown bug living in the intertidal zone on the Atlantic coast. (Included in the Saldidae by Kloet & Hincks, 1964.)

- Over 3mm long or, if about 3mm, then short-winged, with hind margin of wing convex. **24**

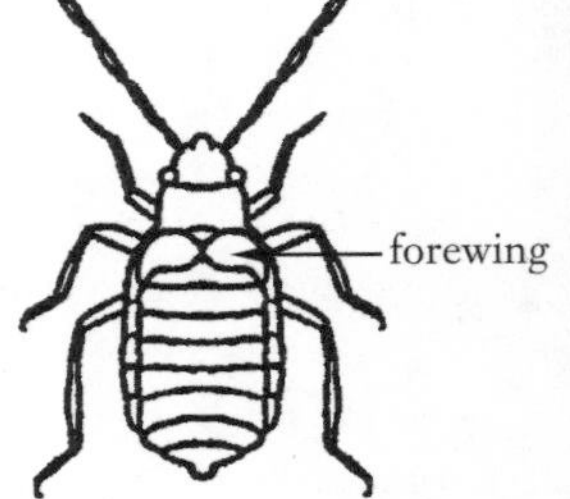

Fig. 42

24(23) Ocelli present (simple eyes on the top of the head, Fig. 43). Look carefully, since they may be small and difficult to see.. **25**

- Ocelli absent. **36**

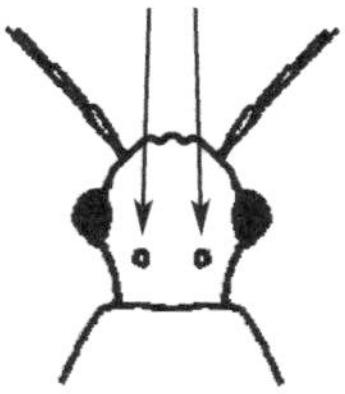

ocelli

Fig. 43

25(24) Rostrum with 3 segments (Fig. 44) **26**

Fig. 44

- Rostrum with 4 segments (Figs 45, 46) **29**

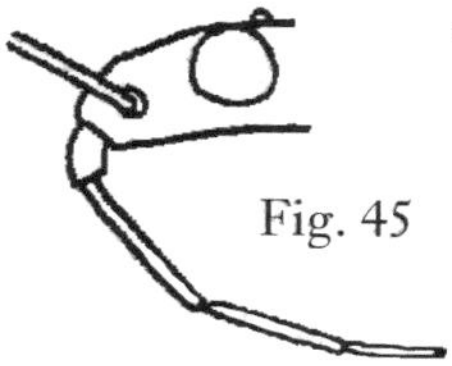

Fig. 45 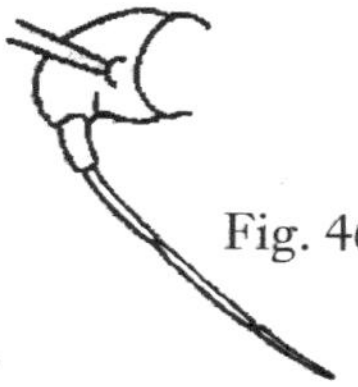 Fig. 46

26(25) Legs and antennae long and thin (Fig. 47), hind legs and antennae much longer than three times the width of the body ....**Reduviidae** (part)
Genus *Empicoris* Wolff, 3 species. 4.5–7mm long. Rostrum stout and curved with the basal end of the 2nd segment well over twice as broad as the 2nd segment of the antenna.

- More substantial bugs (Figs 48-50), with hind legs and antennae not more than three times the width of the body **27**

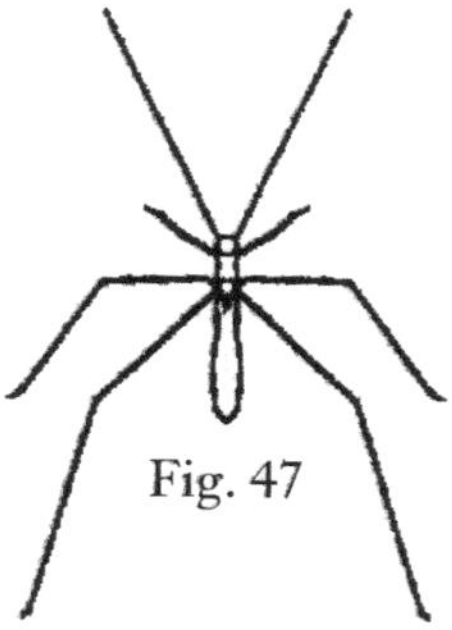

Fig. 47

27(26) Bugs 9–18mm long, stoutly built (Fig. 48). Rostrum curved, with width of the basal end of the second segment well over twice that of the second segment of the antenna. .... **Reduviidae** (part)
ASSASSIN BUGS. 3 genera, 4 species. Predatory bugs feeding mainly on other arthropods.

- 6mm or less. Rostrum with the basal end of the 2nd segment at most twice the width of the 2nd segment of the antenna. **28**

Fig. 48

28(27) Forewing with cuneus and costal break (Fig. 49). Eyes small. If short-winged, then under 2.5mm long. .... **Anthocoridae**

15 genera, 35 species. Feed on small and usually immobile invertebrates. 2-5mm long.

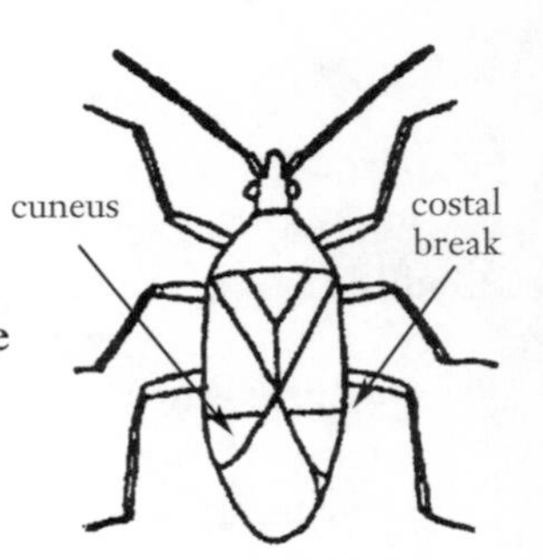

Fig. 49

\- Forewings without cuneus or costal break (Fig. 50). If short-winged, then at least 3mm long. Eyes large, sometimes with a sinuate (wavy) inner margin....... **Saldidae**

8 genera, 22 species. Oval bugs, often black with whitish markings (Fig. 50); some are pale or shiny black. Predatory bugs that hunt on bare ground.

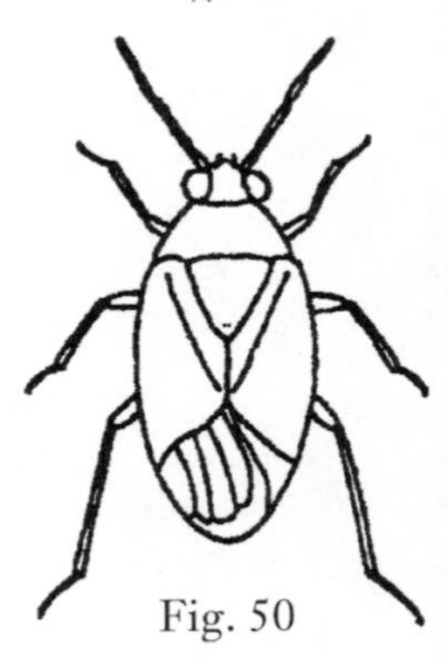

Fig. 50

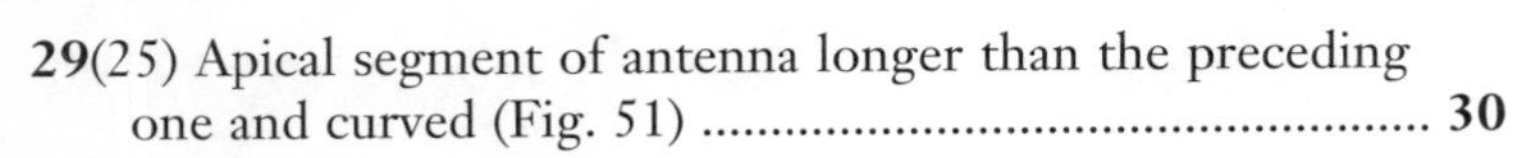

29(25) Apical segment of antenna longer than the preceding one and curved (Fig. 51) **30**

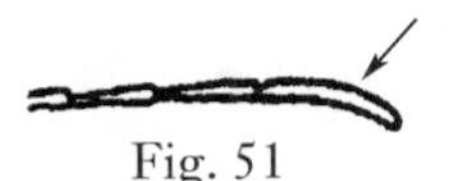

Fig. 51

\- Apical antennal segment usually shorter than preceding one but, if longer, then not curved.............................. **31**

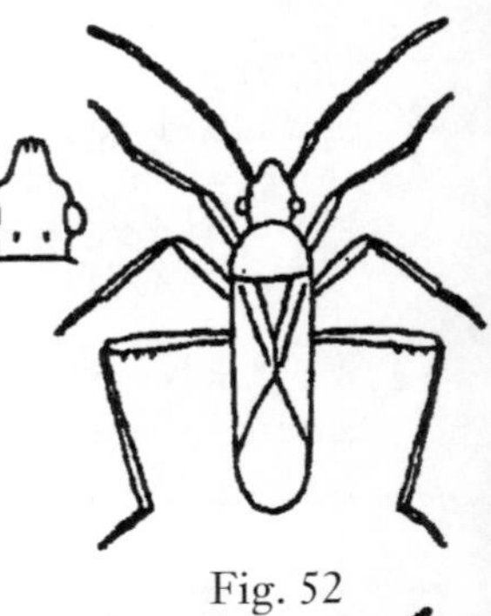

Fig. 52

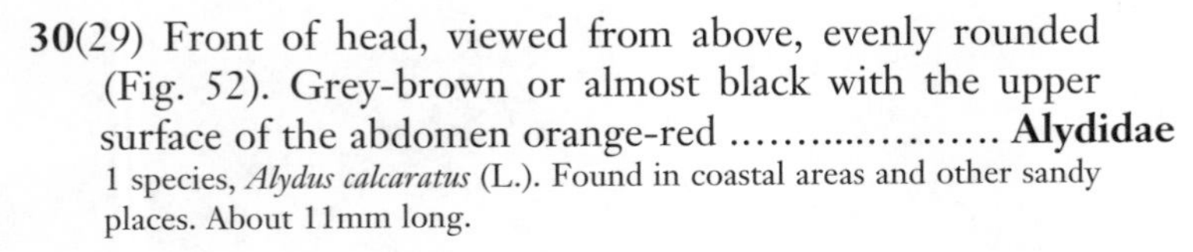

30(29) Front of head, viewed from above, evenly rounded (Fig. 52). Grey-brown or almost black with the upper surface of the abdomen orange-red .... **Alydidae**

1 species, *Alydus calcaratus* (L.). Found in coastal areas and other sandy places. About 11mm long.

\- Front of head, viewed from above, appearing to have two points (Fig. 53). Dark brown bugs with antennae conspicuously banded black and yellow (Fig. 54). .... **Stenocephalidae**

1 genus, *Dicranocephalus* Hahn., 2 species Live on spurges. 8–14mm long.

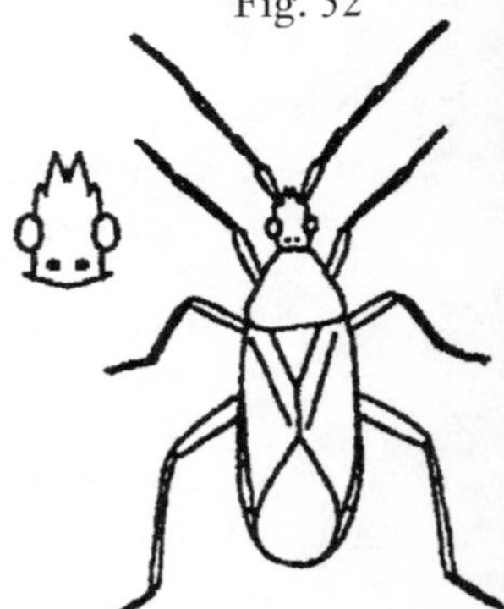

Fig. 53

Fig. 54

31(29) Rostrum stout, curved, and held away from the underside of the body and head at rest. Base of 2nd segment of rostrum about twice as thick as the 2nd segment of the antenna (Fig. 55). If winged, then forewing without cuneus and costal fracture (see couplet 37). .... **Nabidae**

DAMSEL BUGS. Aggressive predators that use the front legs to hold the prey. 7 genera, 13 species. 5.5–10mm long.

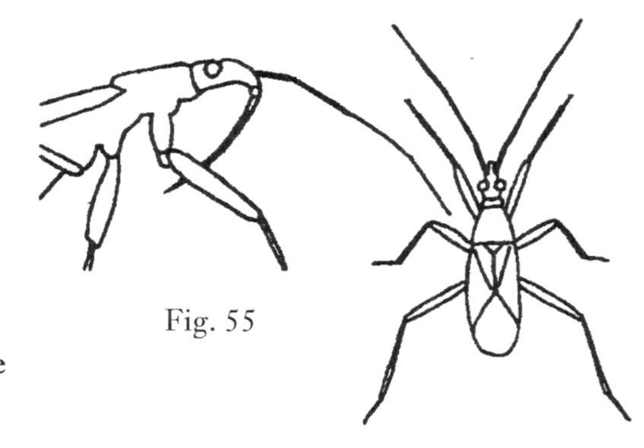

Fig. 55

\- Rostrum more slender, not curved, and held flat against the underside of the head and body at rest. Basal end of 2nd segment of rostrum generally about as thick as the 2nd segment of the antenna, but there are a few exceptions. **32**

32(31) Scent gland clearly visible from the side, located between the middle and hind coxae (Fig. 56). Antennae robust, with the first segment often long and much thicker than the succeeding segments .... **Coreidae**

SQUASH BUGS. 9 genera, 10 species. Fruit or seed feeders. 5–15mm long.

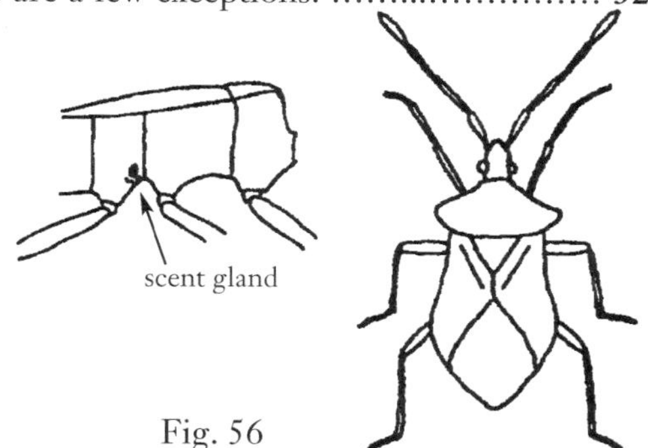

Fig. 56

\- Scent gland, if present, underneath body, and not visible from the side. Antennae usually without a swollen first segment. **33**

33(32) Delicate elongate insects, with hind legs at least 5 times as long as the width of the body (Figs 57 and 58). **34**

\- Broader insects, with hind legs less than three times as long as the width of the body (Figs 59 and 60). **35**

34(33) Last antennal segment (Fig. 57a) expanded and with a dark tip. Femora expanded and usually darkened at tip. .... **Berytidae**

STILT BUGS. Fig. 57b. 4 genera, 9 species. Often found on legumes. 4 –12mm long. Formerly Berytinidae.

- Last antennal segment not swollen, generally yellow. Tips of femora not expanded or darkened. .... **Rhopalidae** (part, Fig. 58)

2 species, *Chorosoma schillingi* (Schummel), 13–16mm long, and *Myrmus myriformismay* (Fallen), 7–9mm, may key out here. Both are found on various kinds of grassland, sand dunes and heaths.

35(33) Membrane of forewing (Fig. 59) with at least 6 veins, usually many more (it is often easier to see veins that are not pigmented by looking at light reflected by the surface). Abdomen coloration often greenish or brownish. .... **Rhopalidae** (part)

5 genera, 9 species. 6–9mm long. Scent gland (between middle and hind coxae) underneath body and only visible clearly from below.

- Membrane of forewing (Fig. 60) with up to 5 longitudinal veins. If short-winged, then abdomen dark brown or black. .... **Lygaeidae**

GROUND BUGS. 34 genera, 71 species. Associated with a wide range of plants and seeds. 2–8mm long.

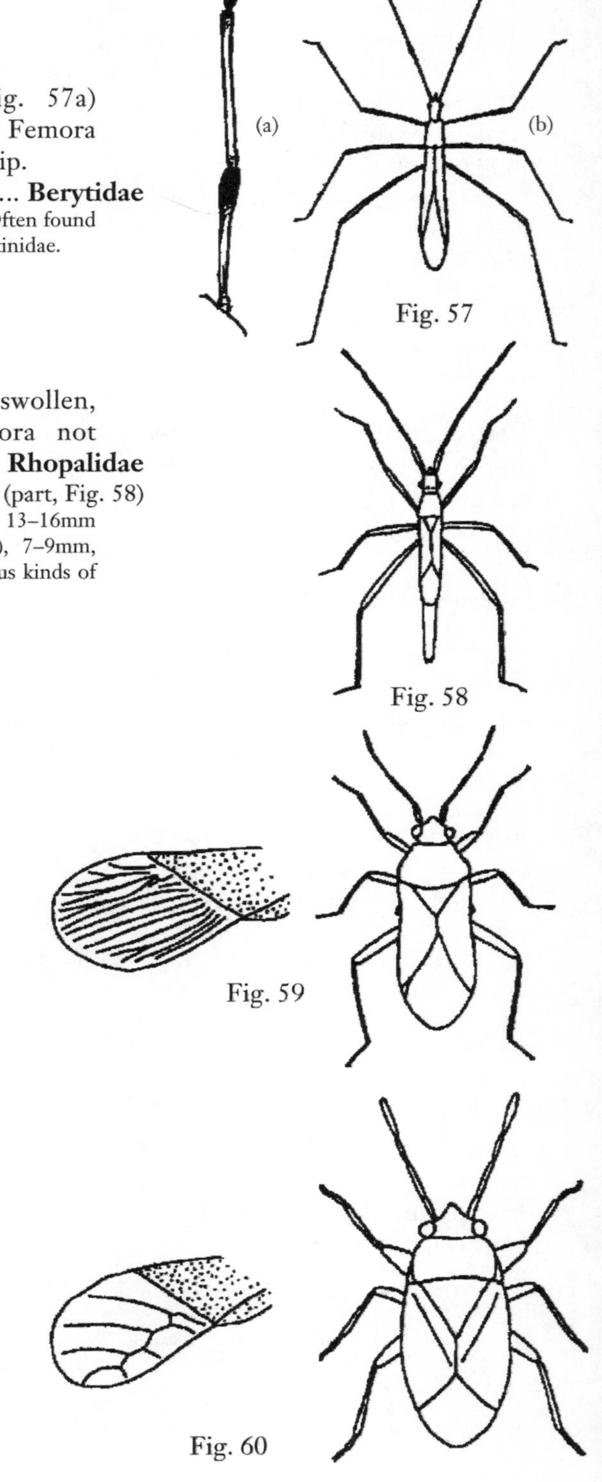

Fig. 57

Fig. 58

Fig. 59

Fig. 60

36(24) 8–11mm long. Scarlet and black bug. Wings usually extending about halfway along the abdomen. .... **Pyrrhocoridae**

FIREBUGS. 1 species, *Pyrrhocoris apterus* (L). Fig. 61. Very rare insect, found on southern and western coasts.

Less than 8mm long, or not scarlet and black. **37**

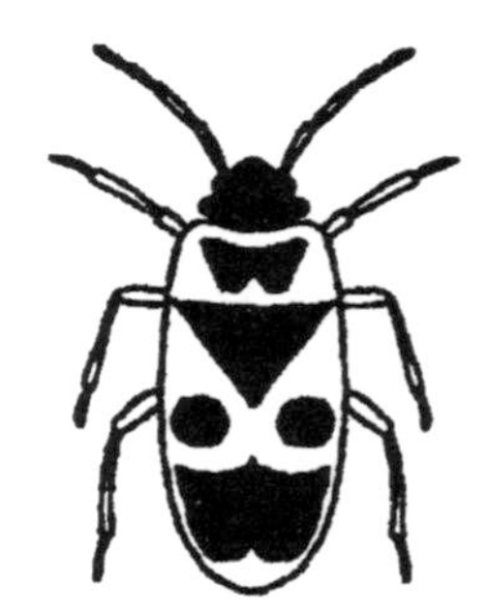

Fig. 61

37(36) Extremely flat, oval, brownish bugs. Wings very short and scale-like (Fig. 62). .... **Cimicidae**

BLOODSUCKING BUGS, (BED BUGS) 3–5mm long. 2 genera, 4 species.

- Delicate, soft-bodied bugs. Forewing with cuneus (Fig. 63), which is the tip of the hardened section of the wing, separated from the rest by a fold running from the costal break .... **Miridae**

MIRID or CAPSID BUGS. Over 80 genera and 200 species. 2–11mm long. A large family of quite varied bugs. Most feed on plants but some are predatory.

Fig. 62

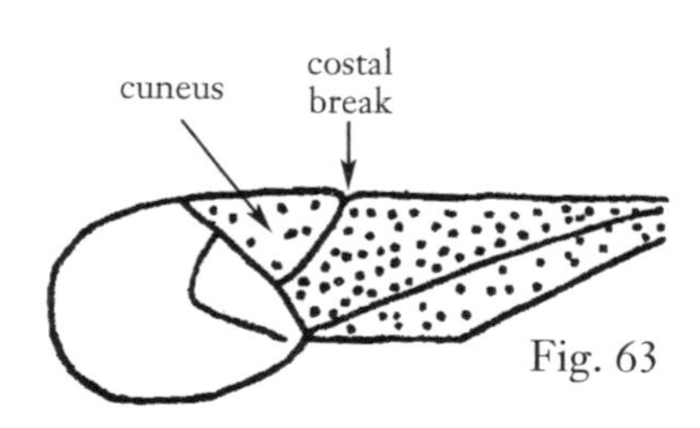

Fig. 63

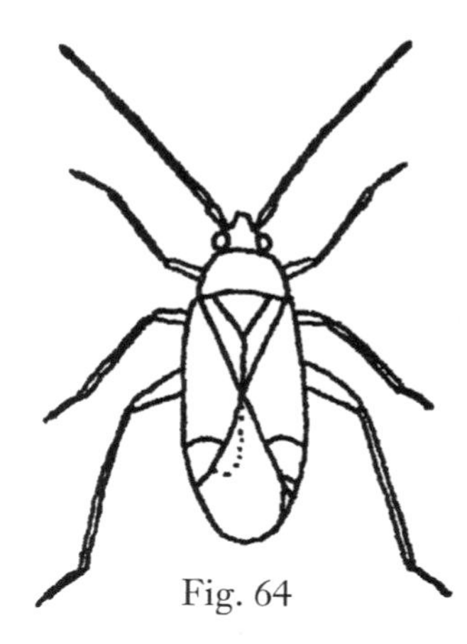

Fig. 64

NOTE: if your bug does not fit either of the descriptions in couplet 37, it may be a specimen in which the ocelli have not developed, or have not been observed. This is known to happen particularly in Nabidae and Lygaeidae, and in the genus *Empicoris* (Reduviidae). None of these has a cuneus. Go back to couplet 25.

KEY TO THE FAMILIES OF THE AUCHENORRHYNCHA

1 Pronotum extended rearwards as a sort of hood over the abdomen and bent so that the front part of the pronotum and the 'top' (dorsal) surface of the head are nearly vertical (Fig. 65). .... **Membracidae**
2 species: *Centrotus cornutus* (L.) which is about 9mm long and *Gargara genistae* (Fab.), about 5mm long.

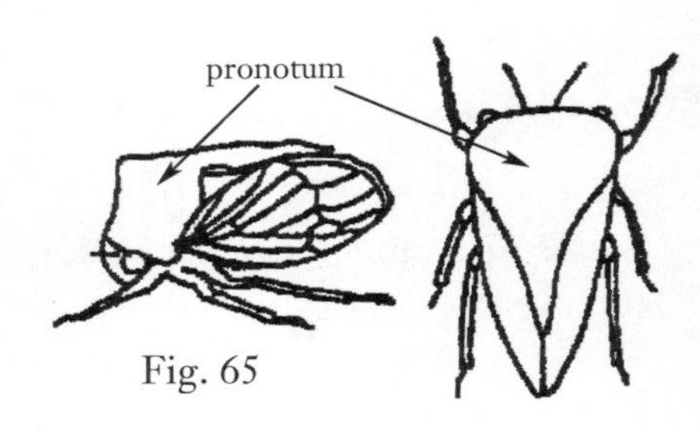

Fig. 65

\- Pronotum not extended in this way. **2**

2(1) Insects about 20mm long. With three ocelli (simple eyes, see Fig. 5 p. 11). .... **Cicadidae**
The only British Cicada, *Cidetta montana* (Scop.) is a very rare insect found only in the New Forest.
THIS IS A LEGALLY PROTECTED INSECT AND, IF FOUND, SHOULD BE RELEASED.

\- Never exceeding 12mm long and usually much less. Two ocelli or none. **3**

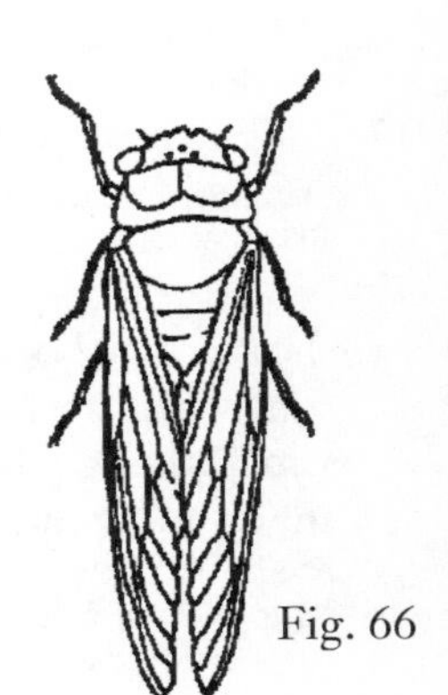
Fig. 66

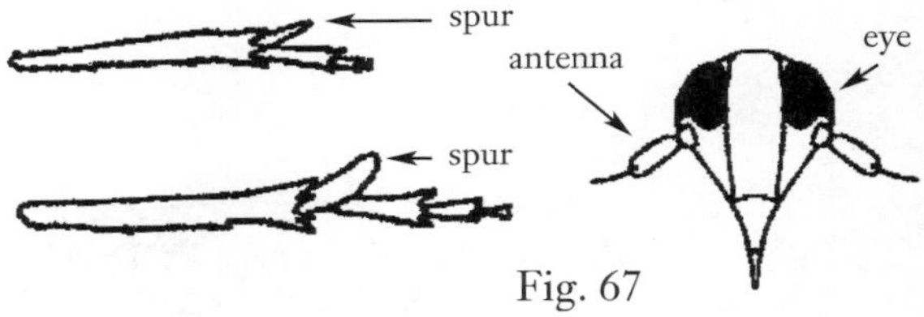

spur
spur
antenna
eye
Fig. 67

3(2) Hind tibia with a large moveable spur (Fig. 67). Antenna inserted near the lower margin of the eye, which is usually hollowed out to accommodate it. .... **Delphacidae**
PLANT HOPPERS. 30 genera, 69 species. 1.5–6mm long.

\- Hind tibia without such a separate spur, although there may be a number of small fixed spines (Fig. 68). Eyes not hollowed out to accommodate the antennae. **4**

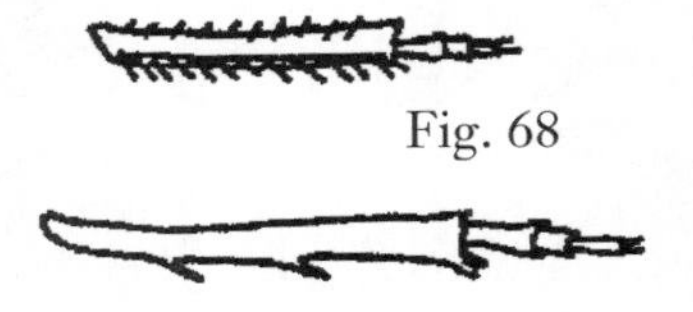
Fig. 68

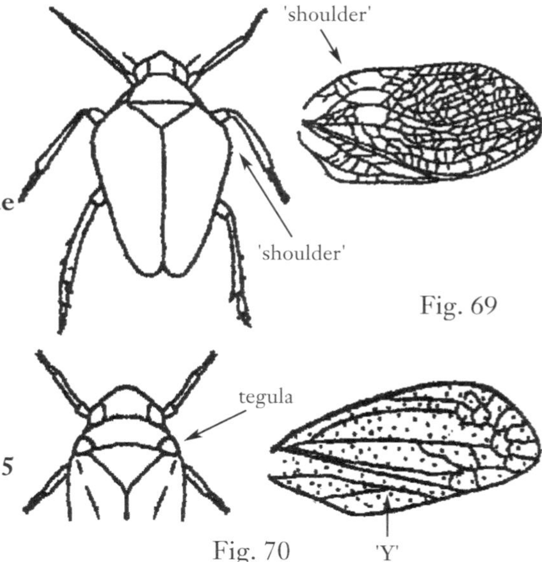

Fig. 69

Fig. 70

4(3) Front wings with leading edge bulging outwards (Fig. 69), giving the bug the appearance of very broad 'shoulders'. Numerous cross-veins present throughout the wing. .... **Issidae**
2 species of the genus *Issus* Fab., 6–7mm long. Found on trees and shrubs, often oak, ivy and whitebeam.

- Leading edge of wing not bulged (Fig. 70) and if extra cross-veins are present, they occur only in the apical (tip) part of the wing (Fig. 70). **5**

5(4) Base of forewing covered by tegula (Fig. 70). Clavus of forewing (the posterior basal part of the wing) with two veins joining to form a 'Y' . **6**

- Tegulae absent. Clavus of wing with parallel veins running to the margin (Fig. 71). Wing veins are obscured by surface sculpting in some frog hoppers (Cercopidae). **7**

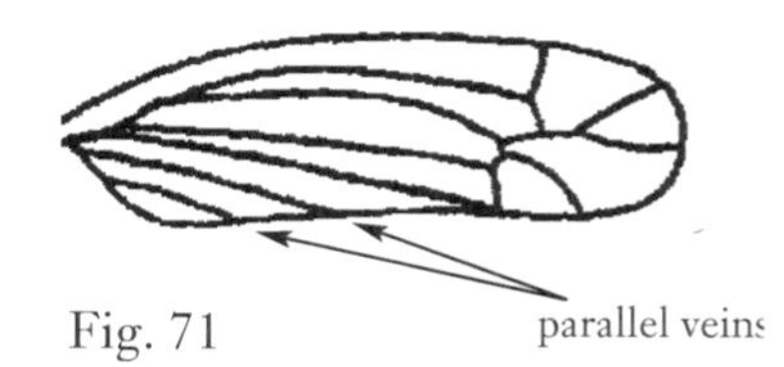

Fig. 71

6(5) Mesonotum with 3 or 5 keels (ridges) (Fig. 72). Forewings clear (hyaline) and membranous, with veins conspicuously raised and standing out clearly.......... **Cixiidae**
4 genera, 11 species. 4–8mm. Found on trees, bushes, grass, and sometimes under stones.

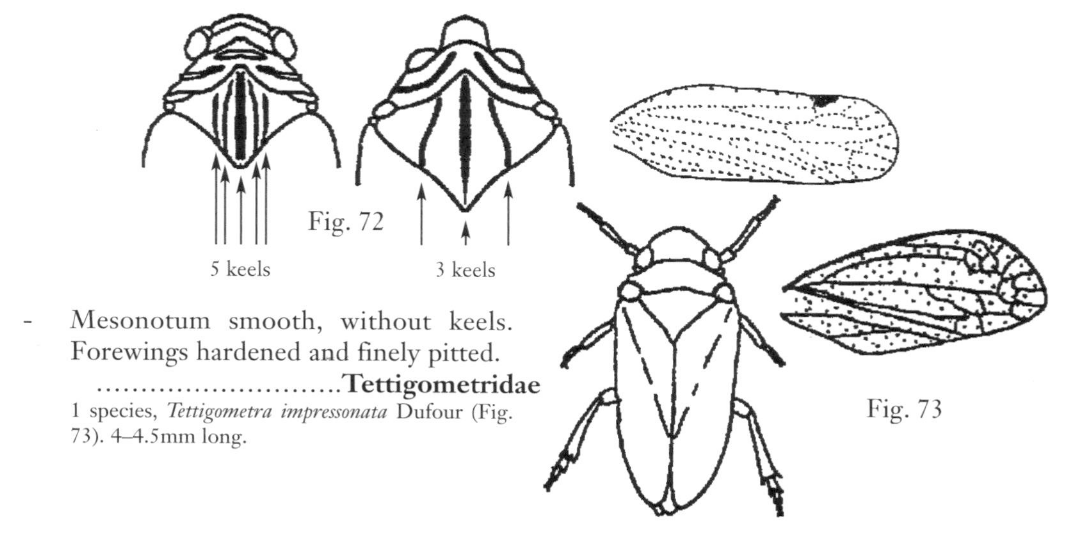

Fig. 72

Fig. 73

- Mesonotum smooth, without keels. Forewings hardened and finely pitted. .... **Tettigometridae**
1 species, *Tettigometra impressonata* Dufour (Fig. 73). 4–4.5mm long.

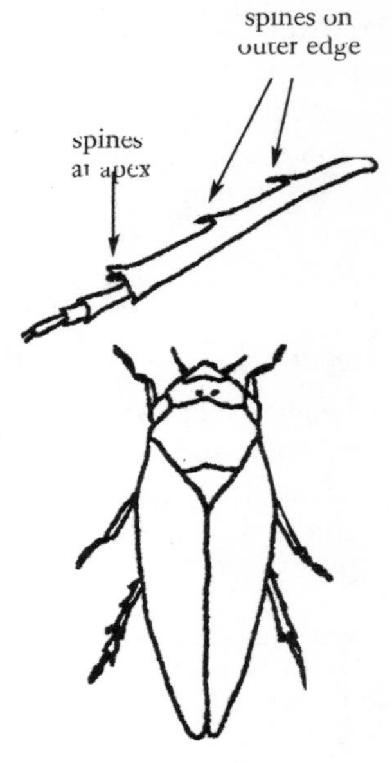

Fig. 74

7(5) Hind tibiae rounded in cross-section, with a few fixed spines at apex, and two strong spines on the outer edge (Fig. 74). ....**Cercopidae**

4 genera, 10 species. 3–11mm. Known generally as FROGHOPPERS or SPITTLE BUGS. The larvae protect themselves by excreting a mass of froth. The sub-family Cercopinae is represented by 1 species, *Cercopus vulnerata* Illiger. This is a striking bug 10mm long, black with six bright red patches. The rest of the Cercopidae comprises the sub-family Aphrophorinae which is raised to a family by some authors.

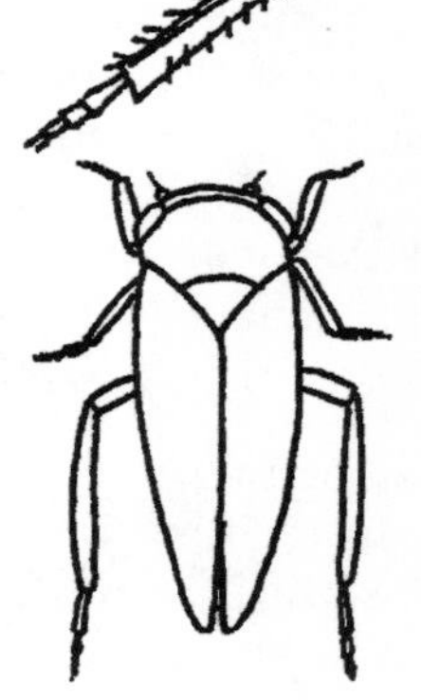

Fig. 75

\- Hind tibiae angular in cross-section, bearing one to three rows of spines (Fig. 75). .... **Cicadellidae**

LEAFHOPPERS. 88 genera, 257 species. 2–11mm long.

KEY TO SUPER-FAMILIES WITHIN THE STERNORRHYNCHA

1 Insect covered with a white waxy powder. Mostly under 2.5mm. Wings of equal size (Fig. 76; left hindwing not visible in this drawing), opaque, and held flat. .... Superfamily **Aleyrodoidea**

WHITEFLIES. 1 family, Aleyrodidae. 10 genera and about 19 species.

\- Insect not covered in white powder. Wings, if present, of unequal size with hindwings smaller than forewings. **2**

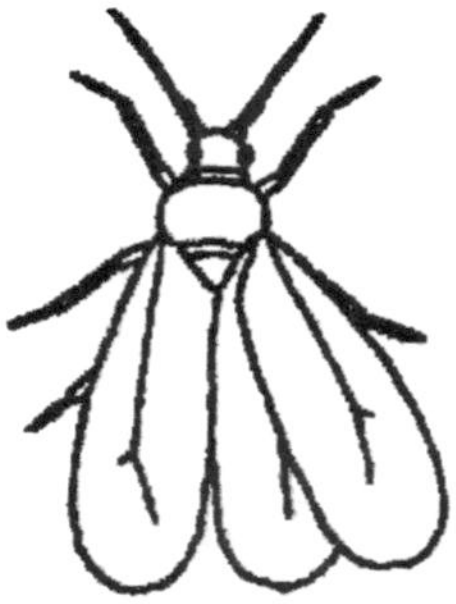

Fig. 76

2(1) Tarsi (feet) with a single claw, or legs absent. .... Superfamily **Coccoidea**

NOTE: this is a difficult character to see on very small insects, and the purpose of checking it is solely to separate out the scale insects (Coccoidea). If you are sure that you do not have a scale insect, go on to couplet 3.

(See Introduction, page 6)

\- Tarsi with a pair of claws. **3**

3(2) Insect fully winged. Forewing with all longitudinal veins originating from a single stem vein in the centre of the wing (Fig. 77). .... Superfamily **Psylloidea** (p. 31)

Known as JUMPING PLANT LICE, PSYLLIDS or SUCKERS. 16 genera, 80 species. 2–3mm long.

\- Wings often absent, but if present, then all longitudinal veins of the forewing originate from a straight vein close to the front edge (Fig. 78). Greenfly, blackfly, aphids or plant lice.................................. **4**

If in doubt, small soft-bodied bugs, often with small tubes (cornicles, Fig. 81) on the abdomen, will key out through this lead.

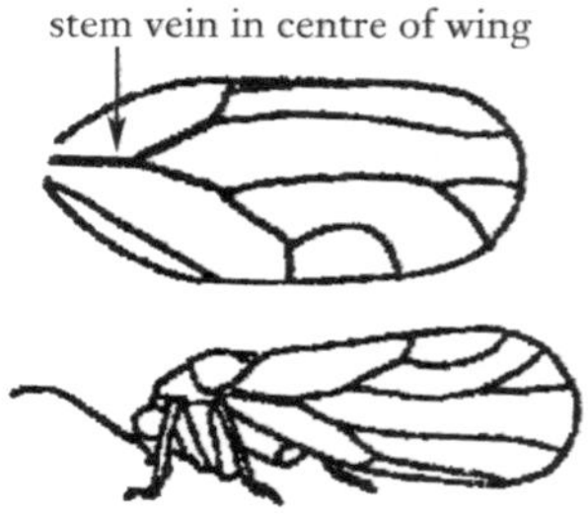

Fig. 77

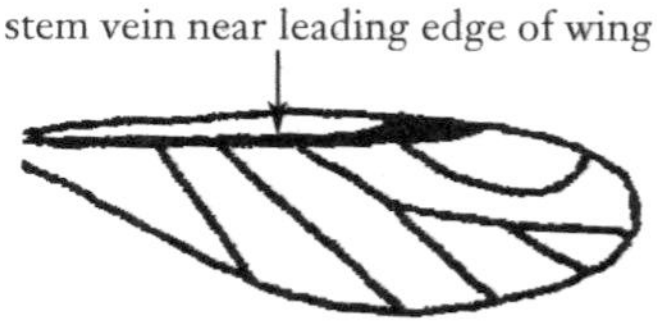

Fig. 78

4(3) Antennae with 3 segments (Fig. 79). Cornicles (see Fig. 81) absent. Eyes with 3 ommatidia (facets). Forewings, if present, with three oblique veins. Cauda (see Fig. 81) short and rounded.
.... Superfamily **Adelgoidea**

Two families. The Adelgidae (2 genera, 10 species), which are sometimes included in the Phylloxeridae, are found only on conifers, where they produce masses of woolly wax or galls. The Phylloxeridae (3 genera, 5 species), are mostly found on oak foliage. The Adelgoidea are closely related to, and were once regarded as part of, the Aphidoidea.

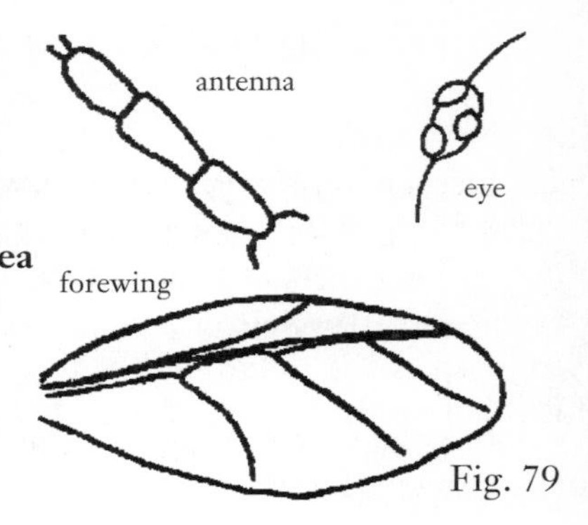

Fig. 79

\- Antennae with 4 to 6 segments, usually long and slender, and often with a terminal process on the last segment (Fig. 80). Eyes, generally, with more than three facets, but a few have only three. Cornicles (Fig. 81) often present but they may be small or absent. Forewings, if present, with four oblique veins. Cauda often long, but see page 35.
................................Superfamily **Aphidoidea** (page 33)

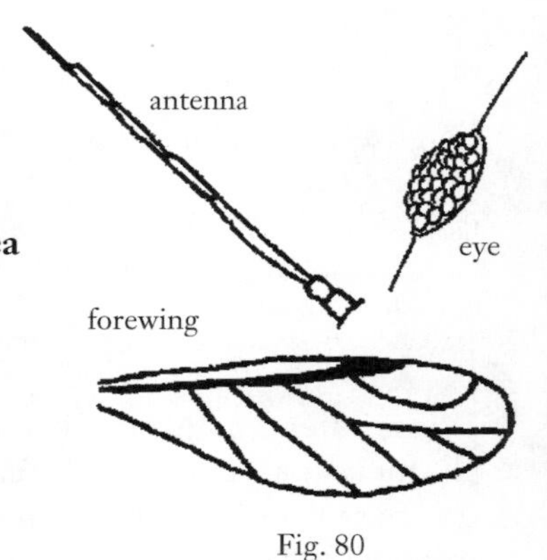

Fig. 80

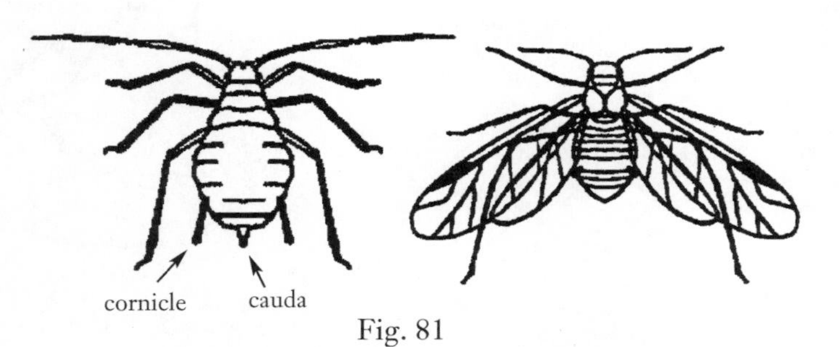

Fig. 81

Guide to the Families of the Superfamily Psylloidea

In the checklist (Kloet & Hincks, 1964), the superfamily Psylloidea consisted of just one family, Psyllidae. By 1979, Hodkinson and White recognised six families, one of which was Psyllidae in a more restricted sense, and Dolling (1991) made further changes. Fortunately, there is an excellent guide to the British Psylliodea (Hodkinson & White, 1979), which bypasses the family concept and goes directly to generic level. However, in the interests of consistency the following family key is provided, which has been derived from the keys of Hodkinson & White (1979) and Dolling (1991).

1 Antennae flattened, bearing long dark hairs (Fig. 82). .... **Homotomidae**

1 species, *Homotoma ficus* (L.), an introduced species associated with fig trees. (Carsidaridae in Hodkinson & White, 1979).

Fig. 82

\- Antennae cylindrical, not flattened and without long dark hairs (Fig. 83). **2**

Fig. 83

2(1) Forewings with veins M and Cu1 originating at the same point on the stem vein (R+M+Cu1) (Fig. 84). .... **Triozidae**

2 genera. *Trichochermes walkeri* Förster lives on buckthorn, and the 17 species of *Trioza* Förster feed on trees, shrubs or herbaceous plants.

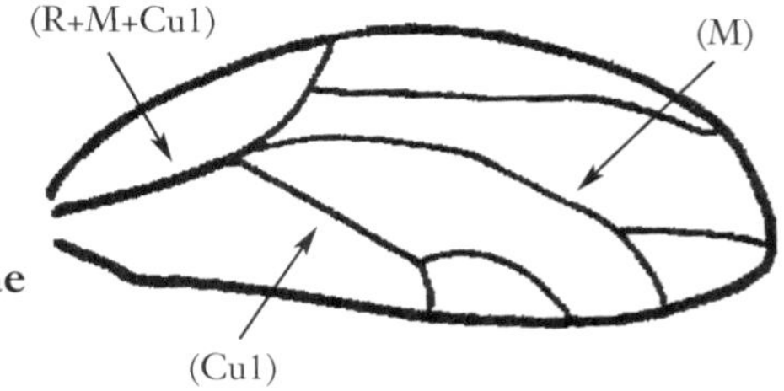

(R+M+Cu1) (M) (Cu1)

Fig. 84

\- Forewing with veins M and Cu1 arising from a common vein (M+Cu1), which runs from the stem vein (R+M+Cu1) (Fig. 85). **3**

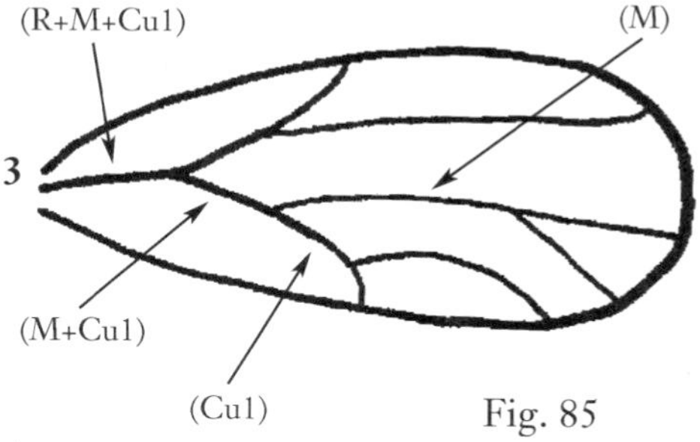

(R+M+Cu1) (M) (M+Cu1) (Cu1)

Fig. 85

3(2) Genal cones present. These projections (Fig. 86) are on each side of the face below the antennae. Viewed from above they look like 'cones' of rather variable shape. **4**

\- Genal cones absent (Fig. 87). **6**

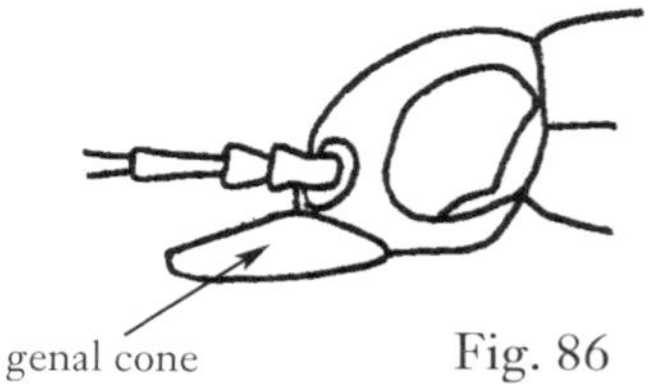

genal cone

Fig. 86

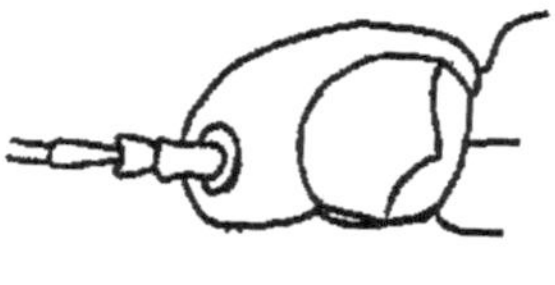

Fig. 87

4(3) Forewing with the length of cell Cu1 almost four times its height. Breeding on *Eucalyptus*.
.... **Spondyliaspidae**
1 species, *Ctenarytaina eucalypti* (Maskell).

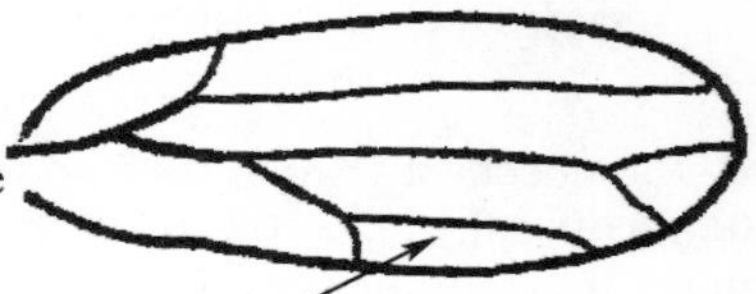

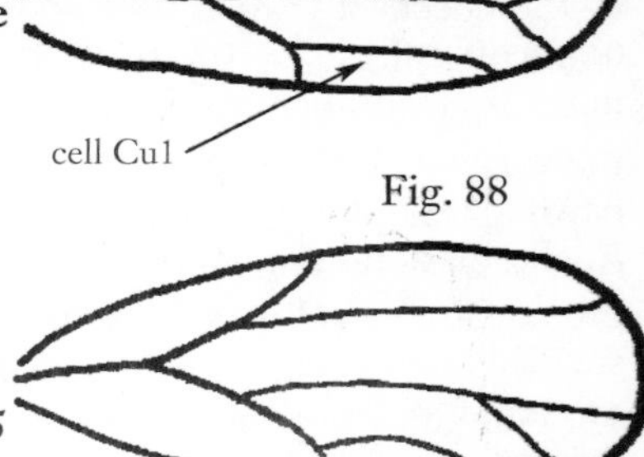

Fig. 88

Forewing with length of cell Cu1 at most three times its height and usually much less. Not breeding on *Eucalyptus* 5

cell Cu1

Fig. 89

5(4) Basal segment of hind tarsus without spines. Antennae shorter than width of head. Breeding on *Cotinus* and *Rhus*.
.... **Calophyidae**
1 species, *Calophya rhois* (Löw).

\- Basal segment of hind tarsus with at least one stout, black spine. Antennae longer than width of head. Not breeding on *Cotinus* or *Rhus*.
.... **Psyllidae** (part)
Subfamily Psyllinae. 6 genera, 16 species.
They feed on dicotyledonous plants.

6(3) Antennae with a swollen second segment that is longer than the third (Fig. 90). .... **Liviidae**
2 species of the genus *Livia* Latreille.
Found on rushes and sedges.

Fig. 90

\- Antennae with the second segment shorter than the third (Fig. 91). .... **Psyllidae** (part)
Subfamily Aphalarinae. 6 genera, 40 species.
Found on trees and shrubs.

Fig. 91

Guide to the Families of Super-Family Aphidoidea

This is a very difficult group of small, soft-bodied insects that are often of great economic importance. The Aphidoidea consists of over 500 British species arranged in an ever increasing number of families, and one should be aware of the changing family concept when using older works. Kloet & Hincks (1964) recognised seven families, including the Adelgidae and Phylloxeridae. Dolling (1991) places these families in the super-family Adelgoidea and recognises 10 other families in the Aphidoidea.

Very small, soft-bodied insects such as aphids tend to shrivel when preserved in a dry state and it is usual to preserve them, initially, in alcohol until they can be prepared and slide mounted for examination at high power under a compound microscope. Such preparation is beyond the scope of this guide, but for those who wish to use the technique, details are given by Disney (1999). It has been decided to limit this guide to what can be achieved with unprepared specimens under a good binocular dissecting microscope, and this precludes the writing of a formal key to aphid families, but it is hoped that the guide which follows will be of some help.

A good starting point is the small book by Roger Blackman (1974), now unfortunately out of print, and a useful tip is that the first move is often to identify the plant on which it was feeding. An ecological approach is often by far the easiest entry to aphid identification since it greatly narrows the range of possibilities. For a more comprehensive treatment, readers are referred to Blackman & Eastop (1985) and Blackman & Eastop (1994). A key to aphid families is given in Dolling (1991), but this requires specimens to be prepared and slide mounted.

In this guide, it is assumed that there are ten families in the Aphidoidea, as recognised by Dolling (1991). Separating them using characters that can be seen on unprepared specimens under a dissecting microscope is hardly possible. The best that can be done is to provide a description of each family as a guide. With care, a large proportion of the aphids that are encountered will be identifiable to family level with a reasonable degree of certainty. These descriptions involve terms used to describe the antenna, and the shape of the cauda (tail) and cornicles (sometimes called siphunculi). These are illustrated in Fig. 92.

Aphididae

This is by far the largest family, with over 350 species. The antennae are long, with the terminal process of the last antennal segment much longer than the basal part of the segment. The cornicles are usually long. If the cauda is long and finger-shaped, the specimen definitely belongs in the Aphididae. However, there are other kinds of cauda in the family. If the cauda is short-triangular or helmet-shaped, then it belongs in the Aphididae provided that the antennae have 4 or 6 segments (not 5) and the anal plate (the plate situated just below the cauda) is smoothly rounded and not divided into two lobes.

Callaphididae

This is the second largest family with about 50 species. The terminal process of the last antennal segment is rather variable in length, and the antenna usually has 6 segments. The cornicles are usually stumpy or elevated on broad cones, but are sometimes long or reduced to pores. The cauda is short, knob-like or rounded, and the anal plate (the plate situated just below the cauda) is often cleft or divided into two lobes.

Lachnidae

The Lachnidae has about 45 species. The terminal process of the last antennal segment is much shorter than half the length of the basal part of the segment. The antennae usually have 6 segments, but some have 5. The cornicles are either pores or are elevated on short cones, and the cauda is broadly rounded. The fourth segment of the rostrum ends in a short, narrow beak. These are mostly large bugs (HAIRY APHIDS).

Pemphigidae

The Pemphigidae has over 40 species. The terminal process of the antenna is shorter than half the length of the basal part of the last segment. Eyes are reduced to three facets. Cornicles are low cones, reduced to pores, or absent. The cauda is broadly rounded. Wax glands are often present on the upper surface of the abdomen (WOOLLY APHIDS).

Chaitophoridae

The Chaitophoridae has about 17 species. The terminal process of the last antennal segment is longer than half the length of the basal part of the segment. The cornicles are stumpy (about as long as broad) and may have reticulate sculpturing, or are sometimes reduced to pores. The cauda is knob-shaped or rounded. The body and legs are covered in long hairs.

Remaining families

Thelaxidae, Anoeciidae, Hormaphididae, Mindaridae and Phloeomyzidae

The remaining families of the Aphidoidea used to be included in Thelaxidae. Anoeciidae has nine species, Hormaphididae has one, Mindaridae has two, Phloeomyzidae has one and Thelaxidae has four: 17 species in all. In all cases the terminal process of the last antennal segment is shorter than half the basal part of the last segment. The cornicles are low cones, pores or absent. The cauda of Mindaridae is the shape of a truncated triangle and the rest are either rounded or knobbed. Hormaphididae, Phloeomyzidae and Thelaxidae are not easy to separate but all have the head and pronotum of winged adults and nymphs fused together so that the eyes appear to be situated at the middle of the sides of what looks like the head. Winged specimens hold their wings flat when at rest, unlike most aphids.

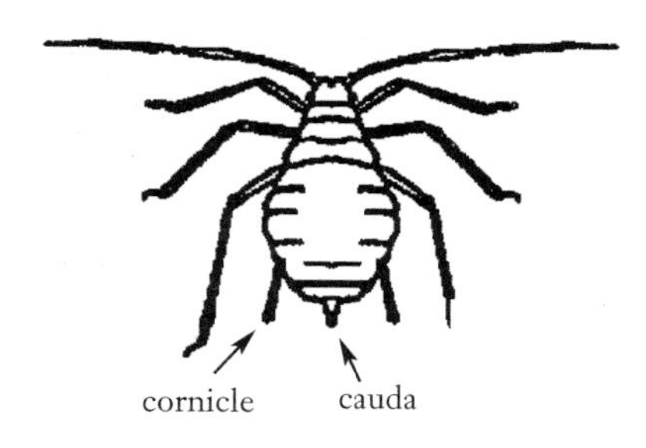

Antennae

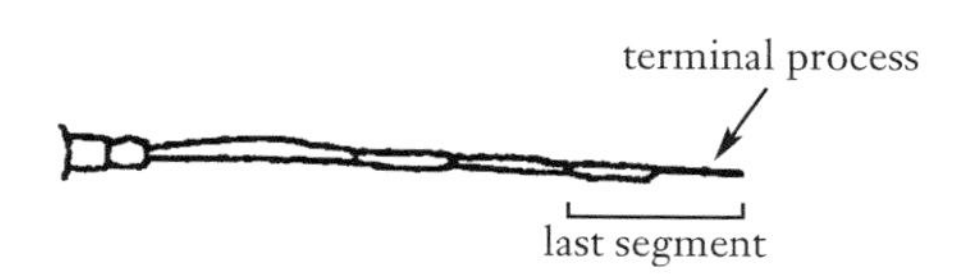

Caudae (tails)

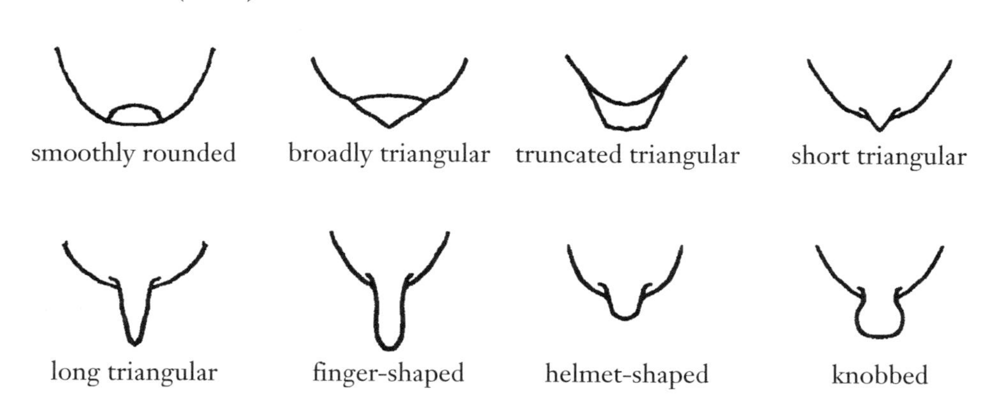

Cornicles (siphunculi)

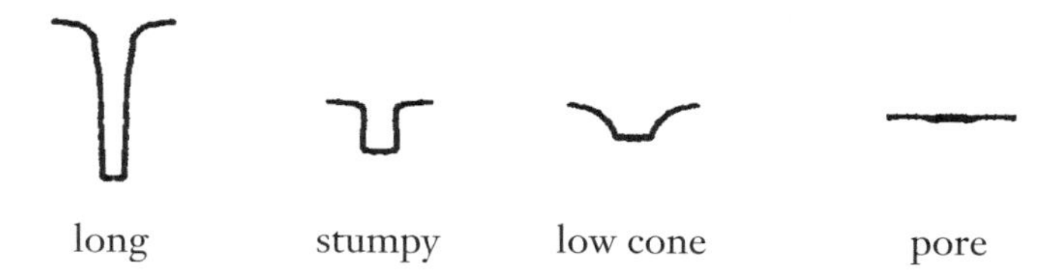

FIG. 92
The terms used in this guide to describe the antennae, caudae and cornicles of aphids